A Complete Guide to Enhancing Safety Culture in the Energy Sector

Niresh Behari

WORLD WIDE
PUBLISHING GROUP

7710-T Cherry Park Dr, Ste 224
Houston, TX 77095
(713) 766-4271

Printed in the United States of America

ISBN: 978-1-68411-864-9

Contents

CHAPTER 1 OVERVIEW

Within the global energy sector, especially in gas processing facilities, there is a history of disasters resulting in catastrophic consequences. In most cases, human and organizational factors contributed to these incidents, which could have been prevented if the right safety processes were established and implemented. That is why it is essential for businesses to become more proactive in investigating the underlying human factors dimensions and to identify and manage critical risks that undermine process safety maturity.

Two catastrophic process safety incidents already occurred in the United States and the United Kingdom. Cullen (1990) found that employees at the Piper Alpha oil rig in Aberdeen, Scotland, inadvertently started the plant while the equipment was still out of commission, which resulted in 200 fatalities. In the United States incident, a Formosa Plastics Corporation employee accidentally turned to the left instead of right and drained the wrong tank, which was in operation, resulting in multiple fatalities. According to the CSB (2006), all the tanks were identical, and this tank contained an explosive mixture.

Process safety culture maturity was defined by HSL (2002) and Khader (2004) as shared corporate values within an organization that influences the attitudes and behaviors of its members. Other safety culture definitions by Khader (2004) include a set of beliefs, norms, attitudes, roles, social and technical practices related to minimizing stakeholder exposure to conditions that are dangerous or injurious. The authors contend that safety culture is a part of the overall culture of the organization and affects the attitudes and beliefs of employees' process safety performance. Both Cooper (2000) and later Erickson (2007) made similar findings in the past, and the literature to date suggests the definition of safety culture has not changed in the last ten years. HSL (2002) also mentions that management is a vital influence of an organization's safety culture and behavior, which also includes:

- Success of safety initiatives
- Incident investigation findings, root causes as well as openness and transparency during investigations
- Compliance to operating procedures, technical, design and operating integrity (popularly known as integrity leadership)
- Undertaking of work-related risks
- Monitoring, correcting and improving process safety critical behaviors
- Health-related consequence management
- Effectiveness and credibility of process safety professionals.

Currently, two governmental agencies in the United Kingdom and the United States address on-site safety system frameworks and process safety deficiencies. The Health Safety Executive (HSE) in the United Kingdom and the OSHA in the United States also developed and currently govern the subject of human factors. HSE (2009) stated that 90 percent of accidents are attributed to some degree of human failure, and the prevention of major accidents depends, to a large extent, upon human reliability at all petrochemical sites. The Occupational Health and Safety Act does not govern human factors or require its inclusion in an organization's safety management system. American and Canadian petrochemical companies, however, recognize the need for institutionalizing human factors throughout the operations environment and are using best practices based on OSHA and HSE recommendations.

HSE (1999b) describes human factors as:

"Human factors refer to environmental, organizational and job factors, and human and individual characteristics, which influence behaviour at work in a way which can affect health and safety."

The three interrelated aspects of the human factors definition are the job, the individual, and the organization. This relationship is illustrated in Figure 1.1 and explained in Table 1.1.

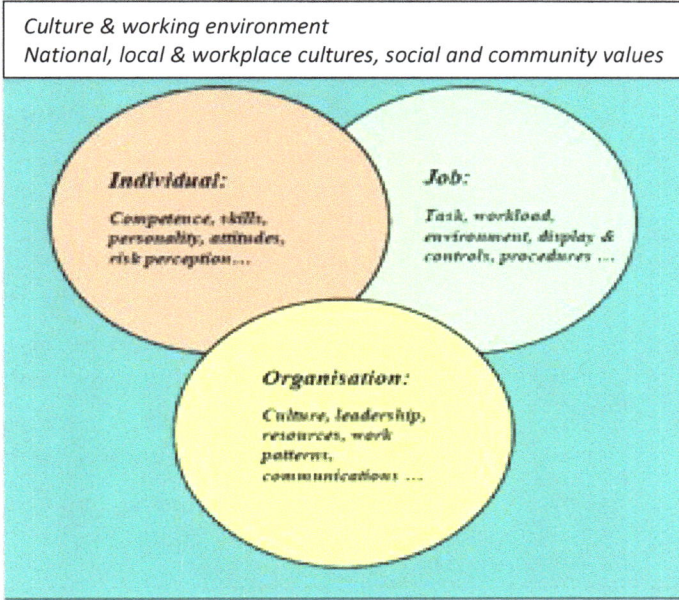

Figure 1.1. Culture and working environment. Data from HSE, 1999b.

Table 1.1. Description on culture and the working environment

Process Safety Culture and Environment	Description
The Job	Nature of task, workload, working environment, design of displays and controls, and role of procedures. Tasks should be de-

	signed in accordance with ergo-nomic principles considering both human limitations and strengths. This includes matching the job to the physical and the mental strengths and limitations of people. Mental aspects would include perceptual, attention span, and decision-making re-quirements.
The Individual	Competence, skills, personality, attitude, and risk perception. In-dividual characteristics influence behavior in complex ways. Some characteristics, such as personal-ity, are fixed; others, such as skills and attitudes, may be changed or enhanced.
The Organization	Work patterns, the culture of the workplace, resources, communi-cations, leadership, and so on. Such factors are often overlooked during the design of jobs but have a significant influence on individ-ual and group behavior.

Source: Data from HSE, 1999b.

The interrelated aspects shown in Figure 1.1 are used to address process safety management system performance improvements where needed. Each topic included in the Human Factors Inspectors toolkit by HSE (2009) helps identify human-related weak spots and incorrect human behavior in process safety and is described in Table 1.2. The toolkit was customized to specifically address the process safety cultural requirements of US OSHA PSM CFR1910 and Canadian CSA Z767-17 (Appendix A) and may be used in any energy generating company in the United States and Canada.

Certain human factors criteria are addressed from on-site process safety audits, while the human and behavior-related issues are investigated by conducting interviews with front-line employees. The audit findings and associated interviews address human factors deficiencies and assist auditors in making recommendations for implementing corrective actions.

Table 1.2. Human Factors Core Topic Toolkit

There are core topics likely to be fundamental to all sites (the Level 1 topics) with a further four topics expected to be common to most sites (Level 2 topics). The remaining topics (Level 3), although important, only will be applicable to selected sites and at certain times in the longer-term business cycle.		
Level 1: Core topics		
1.1	Competence assurance	*Fundamental to good human factors arrangements at all sites*
1.2	Human factors in accident investigation	
1.3	Identifying human failure	
1.4	Reliability and usability of procedures	
Level 2: Common topics		
2.1	Emergency response	*Relevant human factors subjects at most sites*
2.2	Maintenance error	
2.3	Safety critical communications	
2.4	Safety culture	
Level 3: Specific topics		
3.1	Alarm handling and control room design	*Important human factors issues, but only for some sites some of the time*
3.2	Managing fatigue risks	
3.3	Organizational change and transition management	
4	Process Safety Culture	*Assessing the maturity of process safety through organizational cultural norms*

Source: Data from HSE, 2009.

HSE (2010) and PSM OSHA CFR 1910 (2010) describe the technical and behavioral aspects of process safety. The focus is on preventing fires, explosions, and toxic releases, through 14 element standards (see Appendix A, Section 1). An abridged version of OSHA requirements is described in CSA-Z767-17, related to Process Hazard Analysis.

When implementing process safety management, most organizations, including Sasol, Shell, BP, Exxon Mobil, and others, use the HSE Policy, which describes the management's commitment to process safety. Employee participation, change management, safety campaigns, and selected process safety compliance audits address behavioral safety requirements and the degree of safety management system compliance, respectively. The components of human factors are integrated into each process safety element standard and are used to assess the safety maturity culture. These components are in addition to identifying high-level risks and blind spots, preventing an organization from institutionalizing and sustaining process safety.

This description should not be mistaken for a HSE culture survey or assessment evaluation. Typically, these incidents focus on cuts, slips, trips, falls, and pinches and result in low-to-medium injuries or single fatality with a higher frequency of occurrence. Instead, process safety management implementation is designed to mitigate large scale catastrophic incidents that occur infrequently and cause multiple loss of life and major asset or reputation damage due to inadvertent fire, explosion, or toxic releases.

The COMPANY

The organization described in this book is a gas company with 1,100 full-time employees and 1,000 sub-contractor employees and provides gas-to-liquid, ammonia processing, utilities, and fuel dispensing. An assessment was conducted for process safety culture maturity. Process safety implementation occurred for several years before the assessment, although certain process safety elements were used earlier without much emphasis on American Occupational Safety Health Association (OSHA)

Process Safety Management standards or guidelines derived from Canadian Standard CSA Z767. The need for process safety management implementation arose due to numerous incidents in the global petrochemical sector, which resulted in fires, explosions, and toxic releases. Several of the organization's business units experienced numerous process safety incidents resulting in fatalities, equipment damage, and risk of life to the public. The sustainability of process safety in several plants was compromised due to inadequate resource planning, employee complacency, and even unconscious incompetence of employees at all organizational levels. The Dow Chemical Company also requested the organization to implement process safety after several fatalities occurred to its adjacent site during 2006.

Process safety audits showed there was an effective process safety management system in place at the four plants. From in-plant observation audits, however, it was clear employees did not display the correct process safety behaviors when operating equipment. In most instances, employees neglected to maintain equipment, nor did they use any operating procedures. Equally disturbing, adequate operator training was not provided. The audits found that active employee participation and management commitment were lacking in selected plants, and methods were needed to improve the process safety culture.

The launch of the revised permit to work and management of change systems, which form an integral part of process safety, allows employees to work safely and increases awareness of risk coverage before any execution of work. After implementing process safety from 2007-2009, the organization's employees reported feeling burdened by these changes and perceived them as extra work. In fact, the work planning and scheduling process took longer than expected, and researchers observed resistance to change, in part due to inadequate staff and inappropriate work behavior. Other inadequately implemented aspects of process safety were the compilation and execution of standard operating procedures and the updating of process safety information. During the on-site human factors interviews, it was evident older employees retained what they learned and were able to repeat the lessons or operator tasks almost word for word.

On the other hand, newer or junior employees did not fully understand process safety or plant processes. In part, this might have been due to lack of access to operating procedures, inability to make knowledge-based decisions, or absence of adequate training. The human factors interview feedback indicated the risks for these plants increased due to a low level of safety culture maturity and human factors deficiencies, producing repeated process safety incidents.

Another challenge was inadequate communication and interpretation of process safety critical tasks. Different plant labeling methodologies existed throughout the sites across diverse geographic areas. For example, employees might be operating and maintaining the wrong equipment due to the same label used mistakenly for different equipment on adjacent sites, or the equipment might not be labeled in the right series or logical order.

A Safety Maturity Culture

Process safety management is a core business function in the petrochemical sector. It is crucial to entrench lessons learned from past incidents within the organization and other catastrophic global incidents in the hydrocarbon and utilities value chain. Some of the common causes of significant process safety incidents, according to Cullen and Anderson (2005), and Kiddam and Hume (2012), are due to deficient organizational factors and human error applicable to equipment used in the organization. This suggests the necessity and appropriateness for effective operating procedures and investigation of how human error identification and behavior are a contributing factor in preventing such incidents. HSE (1999b) indicates that linkages exist within the organization, people, jobs or tasks, the workplace environment, and operating procedures, which are associated with human factors. Tasks related to a particular job are assessed for risks, human reliability, and appropriate mitigation by conducting a human task error analysis. As a result, even procedural violations are categorized or minimized, and operational areas are identified where additional resources, safety communication, time spent on tasks, or automation are needed.

Identifying safety culture maturity gaps allows an organization to develop strategic goals and objectives for sustaining process safety. When done correctly, this mandate safeguards triple bottom line performance, including profit, environmental integrity, and corporate social responsibility through effective implementation of preventative and corrective risk controls. In this environment, team leadership behaviors and their impact on human factors are assessed to boost safety culture maturity.

Assessing the safety maturity culture through human factors can identify many high-risk process safety deficiencies that exist within the organization, its employees, and the operational job requirements. It's essential to understand the impact of implementing safety culture through human factor dimensions on supporting sustainable process safety management, such as why employees are reluctant to demonstrate the values of process safety in their behaviors and attitudes. Or, how an organization develop effective versus underperforming operations teams, stimulate organizational learning, and identify high-level risks undermining effective process safety sustainability. Determining these answers helps identify the organizational safety maturity stages, provides effective risk-based decision-making, and aids in long-term, robust process safety strategy development and implementation.

Benefits of Improving Process safety

The benefit of process safety maturity assessment indicates how well the existing organizational culture supports process safety sustainability and identifies the gaps needed to create a high-performance process safety culture.

The human factors study is a central theme in process safety culture maturity and effective business performance. The safety culture maturity models selected for this assessment indicate that drivers of key human factors dimensions, including the degree of process safety sustainability, continuous improvement, incident reduction, and reporting effectiveness, are understood by assessing employee team leadership behaviors. The human factors assessment relates employees' behavior and their job roles within the organization to determine safety culture maturity. The benefits of the assessment include identifying risks associated with human factors and providing preventive and corrective actions. The results of a process safety culture assessment help plant management and operational teams develop appropriate safety leadership behaviors that are necessary to sustain process safety, lower incident rates, create a learning culture, and increase organizational safety maturity. Effective organizational learning may reduce or even eliminate process safety incidents without fear or blame during the incident reporting process. In addition, organizational learning stimulates continuous improvement and the ability to comply with growing stringency in OSHA or HSE regulatory compliance and risk reduction.

ASSESSING PROCESS SAFETY CULTURE MATURITY

The need for evaluating process safety organizational maturity culture and assessment through various human factors dimensions encompasses people, their jobs, and the organization. The maturity assessment is accomplished at two levels. The first level is conducting human factors interviews with affected employees and using surveys to identify areas of concerns and risks. The second level is assessing the data for effectiveness and sustainability drivers while also using leadership behavioral patterns to identify potential limiting behaviors that undermine process safety maturity.

Human factors perception surveys, formulated by ERM (2007), and interviews were conducted with employees regarding human factors and process safety culture. The interview questions were developed by the

HSE (2009) human factors questionnaire toolkit and are described in Appendix B. Questions were revised to prevent duplication with those raised in the process safety audit protocols from OSHA PSM CFR 1910, and new questions were included to address the implementation and sustainability of process safety customized for the energy sector. The human factors interview questions used are classified below and were selected strictly based on human-machine interfaces, safety management systems, including organizational communication, learning, and safety culture.

- Maintenance Work
- Emergency Equipment: Maintenance Error
- Control Room
- Alarm Handling
- Process Control System
- Safety System and Safety Culture
- Procedures
- Remote Operations Communications
- Plant Equipment Labeling
- Shift Work Issues

A team of process safety experts conducted interviews using a non-random sampling basis. However, there was an element of bias since the interviewers were responsible for implementing and sustaining process safety and may have had objective or subjective points of view.

The structured interviews were conducted anonymously with operator staff from each department, and managers received interview feedback. A no name and no blame philosophy was encouraged to obtain honest feedback from employees without fear of management retribution. The purpose of the interviews was to compare the safety culture attributes and human factors of the four plants to identify improvement gaps and best practices associated with successful and sustained process safety implementation. Interview results addressing each core topic of human factors and process safety culture dimensions were collected from each plant. The structured questionnaires gathered four types of responses, including opinion, perception, behavior, and attributes. Also, a SWOT (Strengths, Weaknesses, Opportunities, Threats) analysis was completed, and the identification of high-level risks was performed to address deficiencies in the current safety management system based on the perception survey, interviews, and on-site process safety audits.

The human factors survey, which consists of the human factors dimensions shown in Table 1.3, was conducted with all operators and maintenance employees for each department and used to substantiate findings from the interviews, process safety audit results, and the Barret leadership survey (Barret, 1998). The question asked in the survey was:

To what extent, on a scale of one (the least preferred) to five (the most preferred), are the following Human Factors Elements adequately implemented and sustained in your work area?

Table 1.3. Human Factors Perception Survey

Human Factors Elements
Competence and Training
Human Factors & Risk Assessment
Incident Reporting and Investigation
Procedures
Alarm Handling
Maintenance
Behavioral Safety
Safety Critical Communications
Control Room Design & Interfaces
Staffing and Workload
Change Management
Process Safety Management
Supervision
Leadership

Source: Data from ERM, 2007.

The results from the interviews were consolidated to identify common themes using the SWOT analysis, followed by the identification of high-level risks undermining the safety culture maturity throughout the organization. The common themes used to measure the level of safety culture maturity in the organization help to identify the path forward to the next maturity stage.

Additional information was collected from the Barret leadership survey (Barret, 1998) and focused on organizational leadership values, findings, and progress made from safety audits for each plant and process safety incidents severity and reporting frequency. The Barret leadership survey was conducted by a third-party contractor and later assessed for underlying leadership behaviors that were driving or undermining the organization's safety culture maturity. This data was assessed for relevance towards process safety culture and human factors. Process safety incidents, which result in fires, explosions, and releases (FER), were measured for impact and severity using the Fires, Explosions, Releases Severity Index (FER-SI) Tool from CCPS (2009), as a function of the employee population. An internal organizational procedure governs the FER-SI (Restricted, 2009a) to ensure the correct process safety incident classification based on severity, the number of chemicals released, harm to the environment, injury to employees and damage to organizational reputation.

Feedback given to the various plant management teams was monitored for proactive or defensive behaviors and compared to the Barret organizational values results. The overall process safety culture maturity and human factors attributes were consolidated. Common findings and suggestions were translated into a plan of execution to accelerate the level of process safety maturity in the organization.

Recommendations from the human factors interviews and perception surveys were discussed with each plant manager. Team leadership behaviors were qualitatively assessed for commitment towards effective process safety sustainability and management commitment towards implementing recommendations.

Leadership behaviors observed at each plant regarding commitment to implement human factors-related recommendations and Barret leadership survey results were qualitatively assessed to determine process safety maturity culture at each site. Also included in this assessment was the feedback from the human factors interviews and results from employee perception surveys and process safety incident and reporting trends. Three frameworks derived from Eames and Brightling (2012), DuPont (2009), and HSE (2000) were used to assess safety culture maturity relating to organizational learning effectiveness and using a balanced scorecard approach to evaluate individual and team leadership behaviors.

CHAPTER 2 LITERATURE REVIEW

Concept of Process Safety Culture

The results of assessing safety culture and investigating human factors and human error are a good predictor of process safety performance. Knegtering and Pasman (2009) studied safety performance assessment, high-performance organizations, and safety culture maturity throughout the last fifty years. As displayed in Figure 2.1, beginning in the 1970s, safety performance was influenced mainly by human factors, while safety culture maturity originated during the mid-1990s.

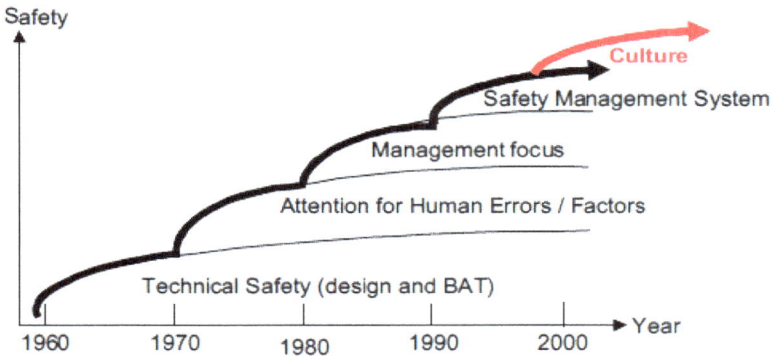

Figure 2.1. Safety performance improvement. Data from Knegtering and Pasman, 2009, 167.

The authors stated that an important lesson learned from the Buncefield oil storage explosion and BP Texas City process safety incidents was the need for a high safety culture maturity supported through effective human factors dimensions to achieve effective safety performance. Some of the common characteristics observed by the authors in process safety accidents in the last decade were:

1. All oil and gas process hazards are known, which suggests "all accidents are preventable and should be seen as a process failure, a cultural failure, and management failure."

2. None of the accidents happened due to a single problem. Instead, they were caused by multiple flaws and deficiencies, which formed a 'bedding' for the accident to happen. The accidents resulted from interrelations between failing system barriers, "unsafe acts'' by front-line operators prior to the incident, inherent safety design flaws, questionable management decisions, and latent preconditions of stress. The interrelations between failing system barriers are called the Swiss Cheese Model (see Figure 2.2).

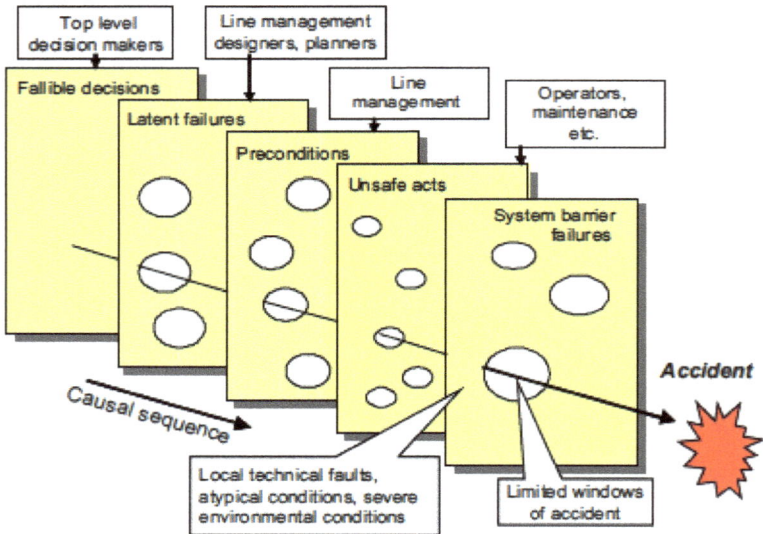

Figure 2.2. Swiss Cheese Model. Data from Knegtering and Pasman, 2009.

1. The 'bedding' that originates in process safety accidents is characterized by management quality and organizational and human factors, which are directly related to an organization's safety culture.

Process safety culture was defined by HSL (2002) and by Khader (2004) as shared corporate values within an organization that influences the attitudes and behaviors of its members. Other safety culture definitions by Khader (2004) include a set of beliefs, norms, attitudes, roles, social and technical practices related to minimizing stakeholder exposure to conditions that are dangerous or injurious. The authors mentioned that safety culture is a part of the overall culture of the organization and affects the attitudes and beliefs of employees' process safety performance. Both Cooper (2000) and Erickson (2007) discovered similar findings and the literature, to date, suggests that the definition of safety culture remains unchanged in the last ten years.

Process safety, according to the OSHA PSM (2010) definition, focuses on the prevention of fires, explosions, and toxic releases using a set of 14 process safety element standards. CSA Z767-17 focuses on the selected process safety risk frameworks, such as Hazard Registers, a hazard, and operability study (HAZOP) and a hazard identification study (HAZID). The remaining aspects of process safety requirements and implementation for Canada are regulated and include Permit to Work and Emergency Response.

The need for a mature process safety culture stemmed from significant accidents, including the fire at Kings Cross, the Chernobyl nuclear radiation explosion, the Piper Alpha explosion, the BP Texas City explosion incident, the BP Oil spill in the Gulf of Mexico and the Fukushima earthquake, which resulted in a nuclear accident. The recommendation made by Fennel (1998) regarding the Kings Cross fire stated that *"a cultural change in management is required throughout the organization."* According to the Piper Alpha inquiry, Cullen (1990) stated that *"...it is essential to create a corporate atmosphere or culture in which safety is understood to be and is accepted as the number one priority."* More recently, in the BP Texas City incident, Baker (2007) highlighted the importance that *"a process safety culture survey be conducted among the workforce at BP's U.S. refineries."* The organization described in this book also experienced numerous process safety incidents with varying severity consequences, and management embarked on fully implementing the process safety management system with the aim of continuous improvement.

Earlier work done by HSE (1999a) references a Business Excellence Model that can be used as a 'vehicle' to drive process safety culture in an organization, as seen in Figure 2.3.

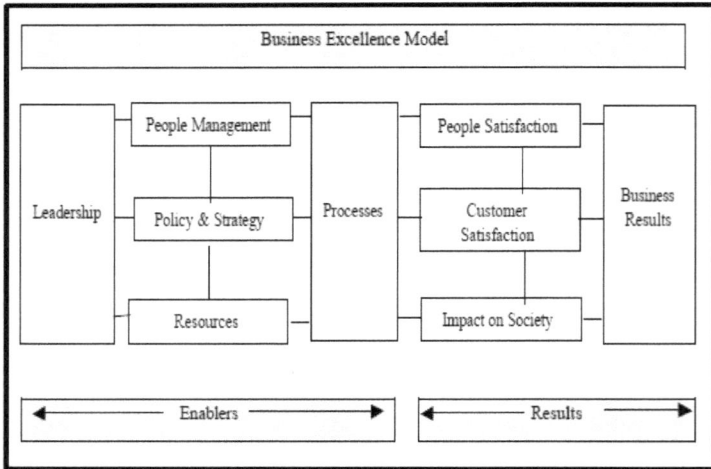

Figure 2.3. Business Excellence Model Supporting Safety Culture. Data from HSE, 1999a.

The authors specify that the Business Excellence Model supports the following attributes:

A. Commitment by senior management to achieve high safety standards demonstrated by communication, consistent decision-making, reward and approval systems, the allocation of resources for employee training and a caring management attitude.
B. Effective communication process built upon trust, openness and mutual respect;
C. Shared view of risks and standards of acceptable behavior that is communicated and maintained;

D. Encourage organizational learning experiences;

E. Health and safety controls and provide governance within a non-confrontational and employee participative context regarding adherence to safety procedures.

The Business Excellence Model was adopted in the organization and encompassed several support mechanisms. An organizational PSM policy charter described high-level management commitment and the process safety leadership required to implement and sustain process safety. A matrix reporting structure was introduced and implemented with employee roles and responsibilities, including process safety awareness as prescribed by CCPS (1994). The organization appointed a project manager to manage the process safety implementation phase and was supported by process safety subject matter experts. A process safety workgroup committee and process safety steering committee were created to track the implementation progress made, the resource requirements, and the path forward as per OSHA PSM requirements.

To achieve business results, the organization conducted internal and third-party audits to measure implementation results and compliance requirements towards safety procedures. Assessments using audit findings, recommendations, and numerical scores were used to monitor progress made and recognize room for improvement. The gaps, detected from experience according to the Business Excellence Model, were in business leadership enablement, people satisfaction, stakeholder satisfaction, and the impact on society. The people and stakeholder focus were addressed with the implementation of human factors assessment, while business leadership enablement was assessed using team leadership Barret survey results and integration with process safety maturity models as described below.

Concepts of process safety culture and maturity include three models that can be interrelated and categorized in five maturity stages. The DuPont Bradley Curve (Figure 2.4) describes different leadership styles that are associated with the five stages of organizational safety maturity from HSE (2000). The more recent Eames and Brightling (2012) model emphasizes five incident reporting and organizational safety learning maturity stages that are correlated easily with the remaining models. The Business Excellence Model, which explains how process safety should be embedded and sustained as well as how to measure business performance, was implemented previously in the organization. However, this model is unable to measure effectively process safety maturity or identify gaps in process safety behaviors that undermine organizational learning or sustainable process safety performance. The human factors perception survey and interview assessment, together with process safety implementation and sustainability progress assessing incident reporting effectiveness and team leadership behaviors, are used to predict organizational safety maturity by using the three models described earlier. Both HSE (2000) and ERM (2008) used a safety cultural maturity model consisting of five levels of maturity (see Figure 2.5).

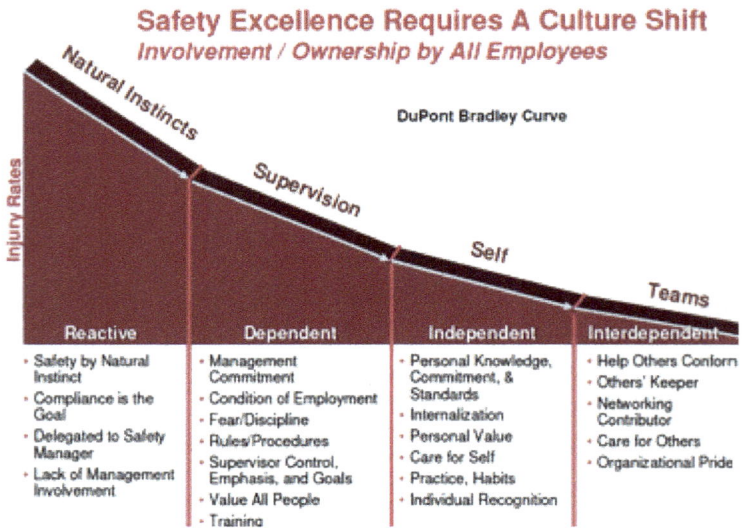

Safety Excellence Requires A Culture Shift
Involvement / Ownership by All Employees

Natural Instincts

DuPont Bradley Curve

Supervision

Self

Teams

Injury Rates

Reactive	Dependent	Independent	Interdependent
• Safety by Natural Instinct • Compliance is the Goal • Delegated to Safety Manager • Lack of Management Involvement	• Management Commitment • Condition of Employment • Fear/Discipline • Rules/Procedures • Supervisor Control, Emphasis, and Goals • Value All People • Training	• Personal Knowledge, Commitment, & Standards • Internalization • Personal Value • Care for Self • Practice, Habits • Individual Recognition	• Help Others Conform • Others' Keeper • Networking Contributor • Care for Others • Organizational Pride

Figure 2.4. DuPont Bradley Curve. Data from DuPont, 2009.

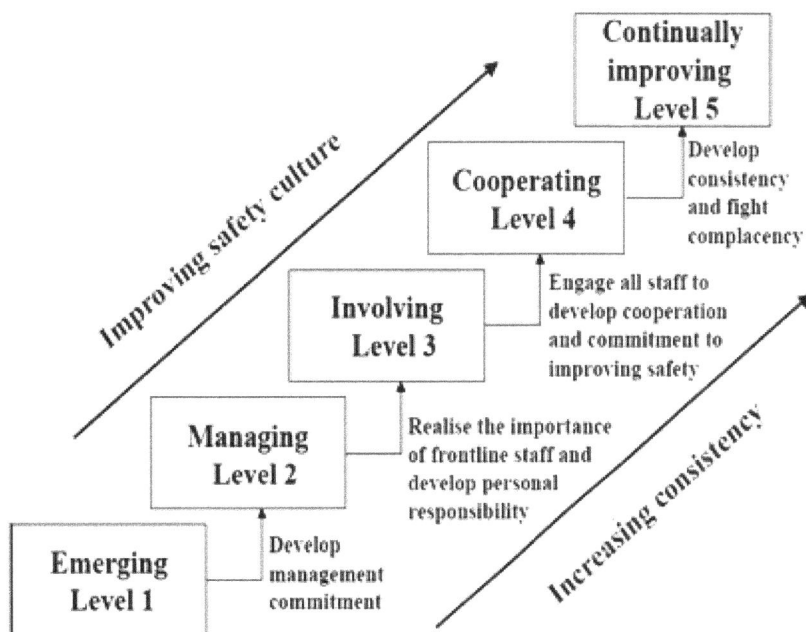

Figure 2.5. Safety Cultural Maturity Levels. Data from ERM, 2008, and HSE, 2000.

Safety culture is a dimension of human factors associated with human behavior and value-driven leadership, which is based on increasing organizational safety maturity. The three safety maturity models described previously, and the human factors assessment are used to identify and assess the existing process safety maturity scale for the organization, and the recommendations provided will enable the fast-tracking of the cultural maturity process. The safety maturity models are supported by a composite view regarding organizational and personal factors, as described in Figure 2.6.

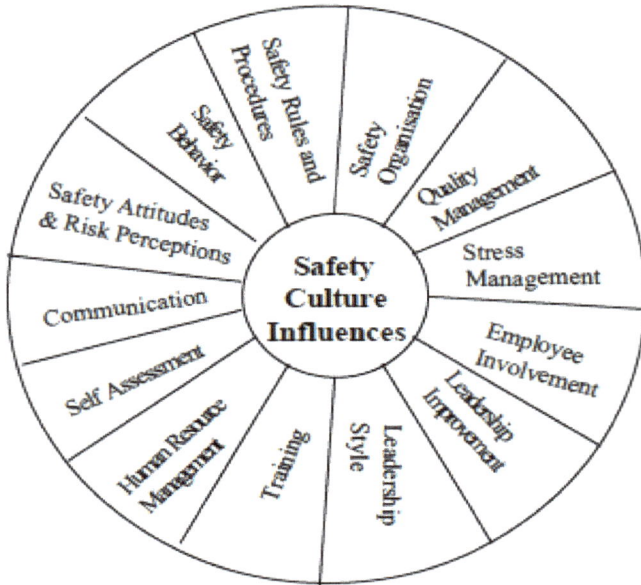

Figure 2.6. Safety Culture Influences. Data from HSE, 1999a.

The human factors employee perception survey, interview assessments, process safety audit results, process safety incidents, and team leadership styles evaluated using the Barret survey will explore safety culture influences. Leadership styles, employee engagement, and behaviors towards process safety will be assessed to identify factors that either undermine or encourage process safety effectiveness and sustainability with the purpose of increasing safety maturity.

Human Factors applied to Process Safety interfaces

Baybutt (1997) stated that people are key elements in any process and that no step in the process lifecycle exists without human involvement and human error. The number of process safety incidents, such as Pipe Alpha, Feyzin, and Flixborough, all are attributed to human error. Baybutt divides human factors into three broad categories: human error analysis, human factors engineering, and human reliability analysis. Both HRA (2008) and Baybutt (1997) subdivide human error analysis into process checklists, task safety analysis, and task error analysis, while human factors engineering consists of engineering reviews and evaluations. Five methods for task safety and error analysis, culminating towards risk identification, were devised by Dalijono, Lowe, and Loher (P332-P335, 2005). These include task data collection (to learn from what you see), task description methods (to link available information and operations), task behavior assessment (to identify what can go wrong), task requirements, and evaluation methods (a form of checklists). Both HRA (2008) and Dalijono et al. (2005) agree that human task error analysis can be devised through human performance failure modes applicable to different tasks. These can be integrated with an organization's existing risk management framework, as applicable to process safety critical equipment.

Baybutt (1997) recommends that the human-machine interface and human-process interface need to be addressed adequately in a human factors assessment, as illustrated in Figures 2.7 and 2.8.

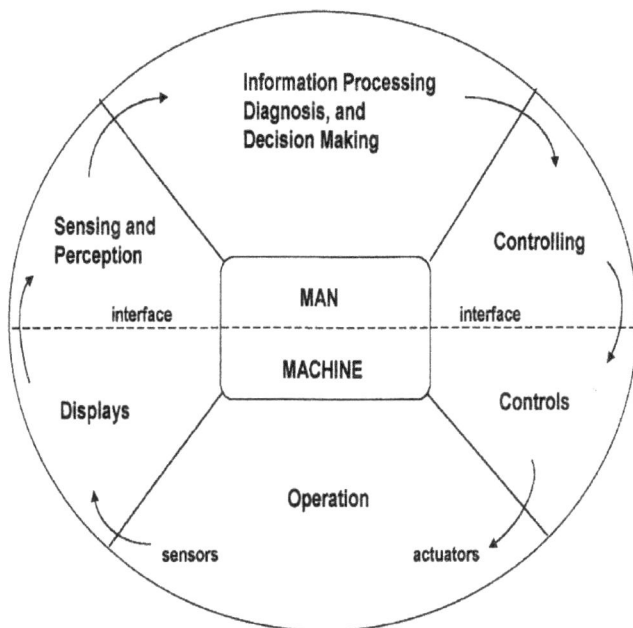

Figure 2.7. Classical Model of Man-Machine Interface. Data from Baybutt, 1997.

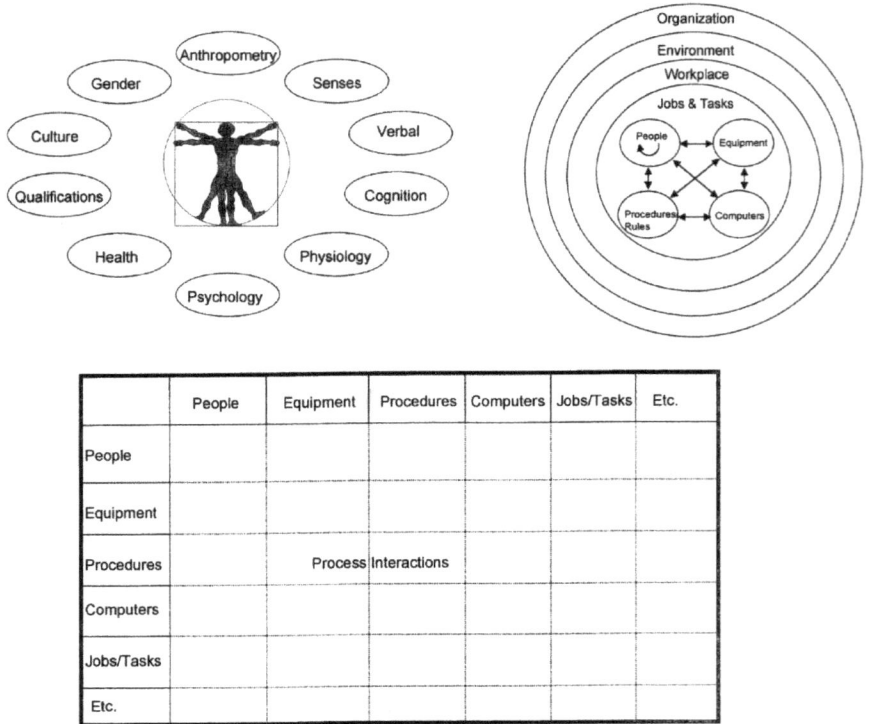

Figure 2.8. Model of a Human-Process Interface. Data from Baybutt, 1997.

Figure 2.8 integrates the human attributes and human reliability with the plant process requirements. These areas are elaborated upon in the human factor's checklist prescribed by HSE (2009). Cullen and Anderson (2005) described the key themes of human factors that are centered on people, jobs, tasks, the workplace, environment, and organization. The authors indicated that human factors are the thread that runs through the safety management system, the organization, and the culture of a site.

HRA (2008), HSE (2009), and Baybutt (1997) describe human error consisting of procedural slips, lapses, mistakes, and violations, which have the potential of causing process safety accidents. The authors suggest that human error minimization should be managed in a structured and proactive way and integrated as part of a safety management system. Some examples of human error from HSE (2009) that relate to performance include the assumption that operators always will be present on the job, treating operators as super-human, assuming operators are well-trained and that they follow procedures or the inappropriate application of techniques. The HSE (2009) checklist includes questions that should be asked to measure the degree of human error minimization and will be addressed in the human factors interview assessment.

Safety culture, human reliability, human process interface, and human error deficiencies are evaluated by the organization using a modified version of the HSE (2009) checklist. The organization edited the checklist for relevance across all plants and to avoid duplication with the process safety implementation and sustainability assessment completed through process safety audits. The interview checklists were completed using a structured interviewing process. The maturity level of process safety was established with the human factors assessment and perception survey, including identifying and managing human factor risks and issues that undermine organizational learning and process safety sustainability performance.

People, Job, Environment and Operating Procedures

Table 2.1 shows the number of accidents in the global petrochemical industries for operating plants (Kiddam and Hume P3: 2012). According to the data, human and organizational factors contribute 20 percent to all accidents. The largest proportion of organizational and human failures are related to storage tanks and piping systems, followed by process vessel accidents.

Table 2.1. Organizational/Human factors attributed to incidents

Accident contributor	Piping system	Storage tank	Reactor	Heat transfer Eq.	Process vessel	Separation Eq.	Total
Human/organizational (a)	41 (18%)	36 (33%)	12 (16%)	12 (16%)	12 (17%)	9 (15%)	122 (20%)
Contamination[a] (b)	17 (7%)	6 (5%)	12 (16%)	11 (15%)	14 (19%)	15 (25%)	75 (12%)
Heat transfer[a] (c)	17 (7%)	10 (9%)	17 (23%)	11 (15%)	8 (11%)	9 (15%)	72 (12%)
Flow related[a] (d)	23 (10%)	15 (14%)	6 (8%)	9 (12%)	10 (14%)	8 (13%)	71 (11%)
Reaction[a] (e)	10 (4%)	3 (3%)	17 (23%)	2 (3%)	12 (17%)	9 (15%)	53 (9%)
Layout[a] (f)	25 (11%)	6 (5%)	1 (1%)	4 (5%)	5 (7%)	3 (5%)	44 (7%)
Fab. const. and inst.[a] (g)	30 (13%)	5 (5%)	2 (3%)	5 (7%)	1 (1%)		43 (7%)
Corrosion[a] (h)	22 (9%)	4 (4%)	3 (4%)	8 (11%)	1 (1%)		38 (6%)
Construction material[a] (i)	19 (8%)	4 (4%)	3 (4%)	8 (11%)	2 (3%)	1 (2%)	37 (6%)
Static electricity[a] (j)	2 (1%)	6 (6%)	2 (2%)	3 (4%)	5 (7%)	3 (5%)	21 (3%)
Mechanical failure[a] (k)	8 (3%)	4 (4%)			2 (3%)	1 (2%)	15 (2%)
External factor (l)	4 (2%)	9 (8%)					13 (2%)
Vibration[a] (m)	8 (3%)			1 (1%)			9 (1%)
Erosion[a] (n)	6 (3%)						6 (1%)
Utility related[a] (o)	2 (1%)					23 (%)	4 (1%)
Total contributors	234 (37%)	108 (17%)	75 (12%)	74 (12%)	72 (12%)	60 (10%)	623
Contributors per accident	2.5	2.2	1.4	2.5	2.1	2.4	2.2

Source: Data from Kiddam and Hume, 2012, 3.

Table 1A in Appendix A indicates that organizational failures contribute to 69 percent of all storage tank failures and are related mainly to poor planning (18 percent) and lack of analysis (16 percent), whereas 31 percent of human shortcomings are attributed to misjudgment and not following procedures. Organizational failures account for 18 percent of piping system accidents, mostly arising from contractor mismanagement (18 percent), work permit violations (12 percent), and ineffective management systems (10 percent). The main contributors of human failures result from inadequate checklists and procedures (25 percent), misjudgment (14 percent), and not following procedures (14 percent). The highest organizational failures (83 percent) are observed with process vessel accidents due to inadequate checklists and procedures (32 percent) and lack of analysis (21 percent). Another (17 percent) of human failures are due mostly to procedural violations (67 percent).

Accidents occurring in turnkey petrochemical projects and applicable to the whole project lifecycle were investigated by (Foord and Gullen,

2006). They believe that changing operator behaviors and equipment design are important both for successful process safety implementation and to reduce incidents. Figure 2.9 shows the primary causes of process safety incidents and include:

- 20% changes after commissioning
- 15% design and implementation
- 15% operation and maintenance, and
- 6% installation and commissioning.

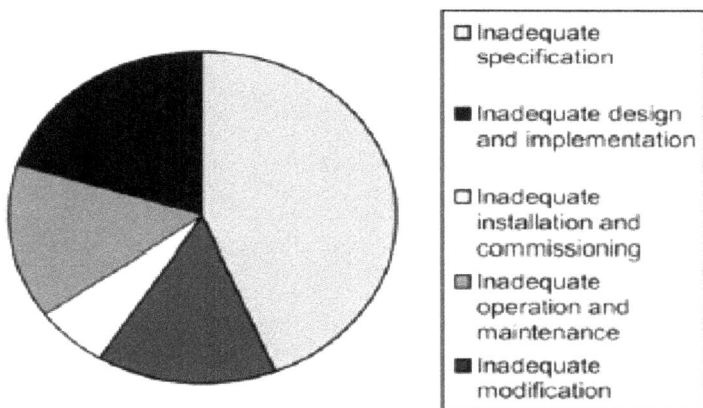

Figure 2.9. Primary Causes of Incidents to Project Lifecycle Phases. Data from Foord and Gullen, 2006.

The authors found that out of 34 incidents, 44 percent had incorrect specifications as a primary cause of the incident, and approximately 30 percent were due to inadequate design, maintenance, and operations. This suggests the need for tighter control towards operational and procedural compliance and effective maintenance and design strategies. Other forms of human factor errors leading to these incidents included design errors, implementation errors and installation and commissioning errors, in addition to operational and maintenance errors.

The analysis presented above indicates that human factors dimensions of ineffective management systems and ignoring operating procedures and self-checklists should be explored in further detail. Some of the challenges faced by the organization are the quality of operator training and the number of tasks attributed to a single operator. Another challenge is when the operator must perform many tasks in rapid succession, especially when performing control room operations and issuing permits that require a high degree of concentration, or work done during shutdowns, emergencies, or when making knowledge-based decisions. Fisk, Ackerman, and Schneider (1987) indicated that combining two tasks can have a different outcome depending on whether the tasks are carried out by controlled or automatic processing, and consistent task performance is directly dependent on practice and learning. Also, the authors suggested the accurate and automatic processing of tasks is strongly reliant on the quality of training and learning provided.

Another nearby plant belonging to the organization by Payne, Xu, Bergman, and Beus (2010) from the Mary Kay O' Connor Process Safety Center at Texas A&M University conducted a process safety culture project. The primary objective of the study was to gather information about standard operating procedures to design an intervention to increase employee knowledge and awareness of the reasons for executing procedures. Although a variety of answers to these questions were gathered, the answers can be organized succinctly into two categories: (1) *Protect the people* and (2) *Protect the equipment/plant*. Protection of the people included the task executor, nearby co-workers, and co-workers upstream or downstream. The protection of equipment focused on proactive ways to avoid damaging equipment.

One of the themes that emerged during the interviews conducted in 2010 that resonated with management teams was the notion that *employees do not know the reasons why procedures exist*. Some of the recommendations made by the authors included employee buy-in on process safety task procedures and fostering a 'big picture' perspective for employees. The attitudes and behavior of employees toward operating procedures are discussed in the human factors perception survey and assessment discussion in Chapter 3.

Kletz (2006) classifies incidents caused by a human operator or procedural errors as slips, mismatches (that are beyond the mental ability of the person), violations, lapses, and mistakes. The need for effective inherent safety design also should be investigated whenever incidents arise. Kletz discourages a blame culture driven by fear and instead advocates organizational learning from these events and "corporate memory." He also states that behavioral safety techniques can be used to minimize human error, although these techniques cannot be used to reduce process safety risks and should not be used as a substitute for effective inherent safety design.

The need to investigate human factors and human errors is due to the number and severity of global process safety incidents, which indicate that several protection layers, according to the Swiss cheese model, have failed. The common root causes identified for process safety incidents include the effectiveness of management decision-making, man-machine interface, human reliability, and human process interfaces, including people, jobs, tasks, the workplace, environment, organizational safety systems, and culture. Organizational and human failures contribute to approximately 20 percent of all process safety incidents (see Table 2.1) and are due mostly to inadequate operating procedures, the lack of planning and analysis, and ineffective operational discipline pertaining to employees and sub-contractors.

The organization experienced numerous repeat process safety incidents and near-misses with low severity consequences. The human factors perceptions and assessment discussed in Chapter 3 identify key human factors risks and issues that currently undermine process safety sustainability performance, safety maturity, organizational learning, operational discipline, and employee trust derived from the human factors survey and SWOT analysis.

CHAPTER 3 HUMAN FACTORS ASSESSMENT

Analysis and Results of Human Factors Survey and Assessment

A human factors assessment and survey were conducted for steam utilities, effluent and disposal, gas-to-liquids, and ammonia plants through an interview with a multi-disciplined team of employees consisting of production employees, maintenance section employees, control room operators, and field operators. ERM (2007) conducted a similar study for the ammonia plant and targeted the above disciplines. A "no-name" and "no blame" approach was used to determine the critical human factors issues that plant employees wanted management to address. The assessment was supported by a human factors survey, where front-line employees quantified their perceptions of different human factors dimensions.

The survey used a well-tested format that provides a semi-quantitative assessment of human factors at the three sites on several dimensions. Respondents were asked to rank each dimension using a 5-point scale, ranging from 1 (reflecting underdeveloped approaches to human factor issues) to 5 (high performance on human factors). The survey, developed by ERM (2007), was conducted with operational/front-line staff in groups of approximately 25 individuals per plant, excluding managers (see Appendix B).

The perception evaluation process and the group interviews attempted to provide a deeper understanding of this assessment to clarify points and any differences in perceptions. The data highlights how human factors are perceived and experienced by different individuals across different functions and layers in the organization. The survey was conducted with maintenance staff, production and engineering supervisors, front-line staff, and control room operator staff. The survey results provided a broad overview of the current human factors on-site and identified areas that would benefit from further improvement (see Figure 3.1).

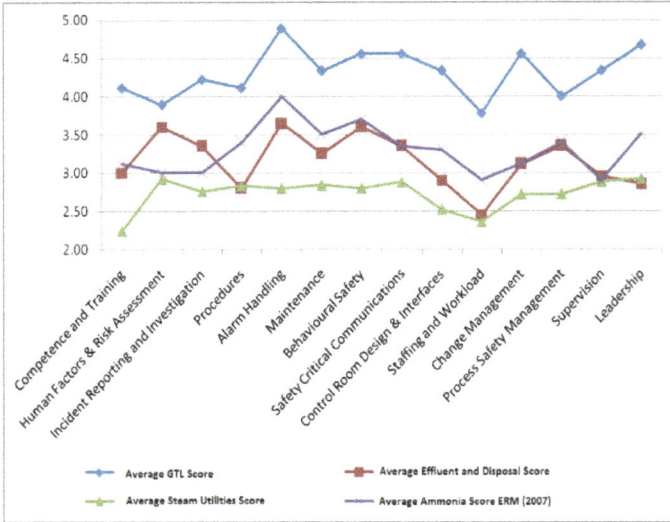

Figure 3.1. Consolidated Human Factors Perception Survey Results.

Figure 3.1 shows the weighted average scores for each human factors dimension based on the four plants. A score of less than 3.5 indicates there is room for improvement, while a score ranging from 4.5 to 5 indicates optimistic perceptions. The operational staff at the gas-to-liquid plant were highly motivated and confident towards living the value of safety. In contrast, steam utilities employees displayed the lowest scores due to organizational restructuring, employee demotivation, and ineffective process safety leadership.

These results show similar trend patterns when compared to the study conducted by ERM (2007). Upward (positive perception) trends are noticeable for Human Factors & Risk Assessment, Alarm Handling, Behavioral Safety, and Change Management. Downward (negative perception) trends are seen for Competence and Training, Procedures, Maintenance,

Safety Critical Communication, Control Room Design & Interfaces, and Staffing and Workload.

Despite the five-year gap between studies, the current work, and that of ERM (2007) demonstrate the human factors perception trends did not change drastically. It also is interesting to note that the ammonia plant was part of another business and only recently was integrated into the organization in 2011. Results from this plant depict a similar trend pattern. Facilitated group interviews with each department clarified perceptions and explained, as part of the qualitative method, aspects of the human factors assessment.

Analysis of Human Factors Dimensions (SWOT Analysis)

The human factors for each dimension were assessed using a consolidated SWOT analysis for all plants based on feedback from the interviews and recent process safety incident findings. A comprehensive human factors assessment and the associated recommendations were compiled for each of the plants. Templates to help energy organizations adopt good practices are in the appendices.

HUMAN FACTORS DIMENSIONS WITH UPWARD TRENDs

Human Factors Risk Assessment

Strengths

Employees in all plants stated they were satisfied regarding the value added by risk assessments and the exercises conducted as part of the safety management system before conducting any work. Risk assessments were taken seriously and performed for site services and process and task risk-related activities through the permit to work process.

Weaknesses and Opportunities

Inattention emerged as a risk since dedicated safety or risk specialists and facilitators were used for each plant. No provision was made to rotate these employees, thus increasing 'risk blindness' in each plant. The blindness of risk was exacerbated due to subjectivity when facilitators were perceived as plant team members instead of providing unbiased inputs to incident investigations, risk assessments, and site inspections.

Threats

Selected safety or risk facilitators who were not perceived as part of the team represented a threat. In this case, they were not trusted easily, and safety recommendations were not implemented fully to lower risks. An 'us versus them' approach between plant operators, supervisors, and safety specialists occurred whenever risk assessments or incident investigations were conducted. The quality of the risk assessment will be compromised if recommendations are not implemented fully, or stakeholders are blind to safety risks.

Alarm Handling and Trips

Strengths

All plants were proactive when responding to critical alarm management. The effluent disposal plant was busy with an alarm management plant upgrade. An alarm trip list was printed monthly for all plants. Also, operators were committed to addressing activated alarms in the shortest possible time.

Weaknesses and Opportunities

Alarm priority lists were scrutinized inadequately to determine any trends, patterns, or room for improvement. Frequent spurious alarms at the effluent and disposal plant need to be eliminated in the new control system upgrade. Experience from past incidents indicated the seriousness of alarm prioritization should be communicated effectively to control room employees. A bias for action should be created whenever critical alarms need to be addressed to avoid plant upsets and discourage complacency.

Threats

Plant and control room operators were unaware of the existence of a bypass trip procedure and incorrectly initiated a change management (MOC) exercise whenever there were alarm trips. It is unclear if trips were authorized, and there was no monitoring of the number of trips made monthly.

Priority alarms could be switched off without supervisor intervention, thus compromising operational discipline and increasing near-miss process safety incidents. Continuous improvement and troubleshooting were undermined since supervisors did not monitor alarm trip list trend patterns.

Behavioral Safety Management

Strengths

Plant and operations employees were enthusiastic about conducting behavioral occupational safety exercises as they were aware of the value of monitoring their colleagues' safety behavior when working in the plant. The safety behavioral program was effective during plant shutdown and was one of the key success factors contributing to a reduction in occupational safety incidents.

Weaknesses and Opportunities

The quality and frequency of behavioral safety observations need to increase as well as the number of people conducting safety observations in the plant. Ineffectiveness was prominent because these observations were made compulsory in office buildings or non-process plant facilities where serious incidents were unlikely to occur or have minor consequences. It is vital for office employees to conduct safety behavior observations in the plant facilities as well.

The behavioral safety program was used exclusively to monitor and address occupational safety. The template used for recording risk behaviors was under a license agreement and could not be amended to include process safety behaviors. Process safety critical task observations and the subsequent revision of standard operating procedures (SOPs) are a compliance requirement described in PSM S1.3 SOP Standard (see Appendix A). However, all plants were unable to comply with this requirement fully due to insufficient resources.

Threats

Too much manpower, as a resource, was wasted in monitoring safety behaviors in non-operational areas. Also, the behavioral safety monitoring, recording, and analyzing process was too long. Employees in non-operational areas perceived the behavioral safety system as not adding value and did not seriously consider addressing at-risk behaviors.

Change Management

Strengths

Change management (or the management of change) is a life-saving organizational behavior, and employees were committed to executing change management for any process or hardware change. In addition, risk assessments, employee training, and updating procedures accompanied any change management initiative.

Weaknesses and Opportunities

Employees struggled to execute organizational and legal changes since they could not fully understand and comply with the MOC procedure. Recent restructuring and retrenchment exercises were conducted without implementing certain aspects of the MOC process and resulted in ineffective organizational changes and process safety risk management. Critical roles and responsibilities were not delineated and communicated to the Human Resources Team during the retrenchment or restructuring exercise, which can lead to loss of corporate and site operational memory

Threats

Major turnkey projects were undertaken without any consideration to change management or if these projects affected nearby sites by compromising utilities or electrical installations. Project managers were unaware that change management was required if they needed to use the site infrastructure, and usually, any new plant or extension increased the overall site risk. Employees were concerned no provision was made to execute change management for new projects or during organizational restructuring. Also, there was a risk of not complying with the new laws or regulations. For example, when the legal requirements changed for vessels under pressure, employees were confused as to how to manage this change since it required additional resources and strategic decision-making. There was confusion whenever a new plant manager was appointed, and requirements associated with legal appointments. This occurred when legal departments were unfamiliar with the employees' roles and responsibilities or how the reporting structure was affected.

Change management for hardware changes was institutionalized in the organization; however, none of the interviewees indicated organizational changes were executed effectively. For example, gas-to-liquid plant employees were relocated to another aging plant section without conducting a change management exercise. They also had concerns about the emergency gathering rooms at the plant since they were unsure if they were adequate and effective enough to mitigate risks. Also, there were no provisions for effective closed-circuit televisions cameras (CCTV) capture throughout the plant.

HUMAN FACTORS DIMENSIONS WITH DOWNWARD TRENDs

Competence and Training

Strengths

Classroom training was well received in the operations environment, and plants included a shift dedicated to addressing training needs. Employees were declared competent for all operational activities, and retraining was provided every three years or earlier, depending on any plant changes or if employees were rotated between jobs. Even contractors' employees received induction training, and their competencies were assessed before being allowed to work on-site.

Weaknesses and Opportunities

Practical on-the-job training was managed ineffectively, and there was an additional workload placed on shift supervisors to fulfill this requirement. Employees struggled to see the value added by theoretical training when it was not supported by practical, on-the-job training. A competency assurance framework on shift supervisors providing on-the-job training had not yet been implemented or audited.

Threats

Interviewees indicate plant operators may be competent to execute tasks based on theoretical training; however, insufficient evaluations of practical training components were assessed. This can increase operational risks and can be misaligned with statutory requirements. These operators also can deviate unconsciously from operating procedures since they were not trained adequately, or they were unable to manage plant upsets.

Standard Operating Procedures (SOP)

Strengths

Operating procedures were written for process safety critical operations and placed on the local intranet. Some dedicated employees compiled training manuals and SOPs. Employees indicated that effective theoretical training was provided, especially at the gas-to-liquid and effluent disposal plant sections.

Weaknesses and Opportunities

Maintenance operations on the equipment included checklists instead of SOPs and process safety incidents (fires, explosions, and releases) occurred due to inadequate checklists or employees not using the checklists to perform maintenance work. For example, during the human factors feedback session with gas-to-liquid plant management, the maintenance manager was hostile towards accepting recommendations and findings originating from the human factors study applicable to SOPs. Observed adversarial behaviors between the maintenance management team and their employees represent a weakness that needs to be addressed. The gas-to-liquid management team perceived that SOPs and checklists should only be reviewed after an accident; however, this type of behavior indicates complacency.

Employees believe trust was maintained by using SOPs to inspect the quality of work done. However, SOPs either did not contain the correct information or were compiled inadequately.

Threats

Procedures in certain areas were not compiled according to the OSHA PSM S1.3 Standard, and procedures were not electronically accessible, including in the steam utilities plant. Control room operators claimed to know the procedure "in their heads." Alternatively, there was an over-reliance on intuition when making knowledge-based decisions during plant upsets. Therefore, it was easier for employees to make mistakes under pressure since they did not have any guidelines and relied on experience to make knowledge-based decisions. Employees believed existing procedures in certain areas did not have troubleshooting guides to aid them in the knowledge-based decision-making process.

Maintenance

Strengths

Employees took responsibility whenever equipment was used beyond normal operations by obtaining the necessary concessions. Equipment was maintained consistently, and management emphasized equipment reliability during their inspections. Maintenance work was prioritized according to the Asset Integrity Process Safety Management System.

Weaknesses and Opportunities

Employees lacked sufficient resources to develop maintenance management systems, compile maintenance strategies, or upload information regarding maintenance frequencies or mean time between failures. Ergonomics were not considered during maintenance task execution, and employees complained about standing on railings, entering cramped spaces, or carrying too many tools and objects while climbing stairs and scaffolds. The identification of safety critical equipment using a recognized framework needs to be established.

Threats

None of the interviewees could identify process safety related to critical equipment; however, employees were concerned about access to equipment. Elevators did not work, and employees stood on guard rails to conduct maintenance operations without being provided with scaffolding. Tools were carried instead of using a maintenance box at elevated platforms due to ineffective operational discipline. Interviewees indicated there were inadequate maintenance resources for sustaining process safety asset integrity.

Safety Critical Communication including Equipment Labeling

Strengths

Safety critical communication was performed using a public announcement (PA) system tested weekly. Surveys were conducted for sound effectiveness, and the need for repair or additional installation of PA systems was addressed. Employees also received training on communicating safety issues with limited success.

Employees responded well within the time limits for reacting to emergencies during safety announcements. The local radio stations were used to convey emergencies. Dedicated emergency contact numbers were available for the police, ambulance services, and road traffic planning.

Weaknesses and Opportunities

The efficiency and maintenance of two-way radios were inadequate at the steam utilities plant. If a control room operator was on site and was unable to communicate to the control room, there was no way of determining whether the operator was in distress. Public announcement systems for safety and emergency communication may be inadequate in high noise zone areas near the steam utilities boilers. Closed-circuit television was used; however, zoom and motion capability were restricted. Control room operators were unable to identify potential incidents at gas-to-liquid, effluent and disposal, and steam utilities.

Equipment labeling can be improved when pumps at the effluent and disposal and steam utilities plants are located nearby, and it is possible to switch off the wrong pump manually. For example, control room operators at the steam utilities plants previously switched off the wrong pumps and were able to recover the process, suggesting the need for pump switches to be installed further apart.

Threats

Safety critical communication could not be heard clearly, especially when maintenance work was conducted at the steam utilities plant. Employees remained confused as to whether the emergency situation would affect them, if they need to evacuate, or if the emergency related to another plant section.

Employee participation was not conducted with control room employees when CCTVs were installed. For example, the cameras at the gas-to-liquid plant were fixed, and any loss of containment could not be seen immediately before a catastrophic incident, adversely affecting operator reaction times. Moreover, control room operators also were appointed to manage on-site emergencies for an aging methanol plant. The control room was centrally located, and longer reaction times were experienced when managing on-site emergencies.

Experience indicated pipe labeling was dissimilar in all sites and created confusion with contractors who worked at different locations. The color-coding of pipes was different across sites, and equipment identification for work to be done under permit to work conditions had different 'touch and tag' procedures across the plants. Two process safety incidents occurred at a nearby hydrocarbon wax production facility and effluent disposal plants from March to October 2012, due to non-compliance with different 'touch and tag' procedures executed by contractors. For instance, the wax facility required two tags per equipment (production and construction) before any work could start, while the effluent disposal plant only needed one. In both incidents, contractors began cutting the wrong pipe, which contained high-pressure hydrogen gas. This nearly caused a catastrophic process safety incident because the pipes could have exploded and resulted in multiple fatalities and equipment damage.

Control Room Design & Interfaces

Strengths

Management and staff recognized the need for upgrading control room design layouts. For example, during the assessment, a control room upgrade was in the process of completion at the effluent disposal plant.

Weaknesses and Opportunities

Outdated control rooms were a weakness. The control room at the gas-to-liquid facility was approximately six years old and considered "state of the art," however, operators struggled to see the visual layout. They made mistakes more frequently due to incorrect "mouse clicks" during navigation since there was no room to enforce calm reflection before making knowledge-based decisions. These operators could not clearly distinguish color fonts on the operator control panel and had not taken color blindness tests. No provision was made on the digital control systems to allow recovery of an operating process during an upset within pre-defined time limits, and there was insufficient time for operators to think critically before making knowledge-based decisions. The control room at the steam stations was outdated and did not have an adequate display layout for the boiler start-up and shut down process. The pumps switches and alarms were not associated to each boiler and were grouped together; thus, the wrong pump was inactivated on several occasions

Threats

Control room layout varied for each plant due to age, and there was no consistency regarding the layout of information or even a uniform alarm management philosophy. Employees at the gas-to-liquid plant struggled to read the control panel due to the color font. A protection layer for confirming key operator inputs only existed at some of the plants, and the alarms were used as a trigger for operational control instead of indicating a process upset. The additional protection layer for standard plant control helped reduce inadvertent mouse clicks, which could trigger unwanted process operations. Alarm logs were never inspected for false alarms or when alarms were canceled without operator intervention, which could lead to a process safety incident.

Staffing and Workload

Strengths

All plants were adequately manned per shift and used an effective shift hand-over communication policy. In addition, standby and overtime were well managed so that employees were not fatigued. Plant managers ensured compliance with overtime policy and labor law legislation.

Weaknesses and Opportunities

Employees were required to perform "multi-tasking" roles, where they felt pressured to execute several tasks without enough time. Control room operators felt their concentration was compromised whenever they needed to execute more than one task in a short amount of time or if they needed to answer phone calls, check emails, write or issue work permits or intermittently leave the control room. For example, gas-to-liquid methanol plant control room employees may need to travel as far as a five-minute distance during an emergency because the control room was not situated nearby the plant. It is unclear whether the current manning level was sufficient to address plant upsets or emergencies. Also, the CCTV camera for this plant was ineffective, and it was unclear whether the control room employees can respond to emergencies on time.

Shift supervisors also felt pressured and could not perform effectively since they needed to provide on-the-job training to junior operators. This was because the training officer provided only theoretical training.

Safety resources need to be more visible in the plant. Plant operators perceived that safety practitioners spend too much time in their offices and not enough time walking through the plant where they could detect and address incorrect safety behaviors.

Threats

Staffing and workload were not optimized effectively. Employees felt overworked or experienced the need to accomplish several tasks in a short amount of time. The gas-to-liquid methanol plant had three control room operators, yet nobody was appointed to man the plant on-site. Similarly, there was only one control room operator in the effluent disposal plant, and two were on-site. An incident of overfilling occurred in the thermal oxidation area during an emergency, and it was discovered that the control room operator could not cope with managing the evacuees as well as control the overfilling of the high Sulphur pitch tank. The operator dismissed the alarm and allowed the tank to overfill. Additionally, the two on-site operators did not adhere to operating procedures for the tank pump filling process.

The interviews indicated work prioritization and planning, especially for on-site and control room operators, were ineffective. There was a discrepancy between the number of staff hours needed and the actual number of staff appointed.

Process Safety Management, Supervision, Leadership and Safety Culture

Strengths

Dedicated maintenance services and supervision were provided for each plant, and employees were not required to attend meetings outside their scope of direct responsibility to save time and safeguard costs. Interviewees were satisfied that maintenance supervisors were effective in articulating organizational strategy and communicating action plans to maintenance employees.

An effective safety management system was implemented and continually updated based on findings or recommendations from auditors and safety consultants. Best industry practices were implemented in each department. The executive management committee and plant managers conducted weekly safety inspections. For example, employees at the steam utilities, effluent disposal, and gas-to-liquid plants felt confident and had no fear of speaking to management regarding safety issues. The effluent disposal management team also highlighted reliability and safety issues during their inspections. There was a good balance between the occupational and process safety inspections conducted by management, and safety issues were given a higher priority than meeting production targets.

Employees were aware of and understood the need for process safety management. They were committed to complying with the permit to work, energy isolation, and management of change processes. Also, addressing actions that arose from risk assessments played an integral part of employee roles and responsibilities. Employees had the right attitude and behaviors when responding to emergency exercises and did not delay in evacuating to a nearby emergency gathering room.

Leading and lagging balanced scorecard indicators were monitored as a "health check," and employees were assigned and committed to completing actions arising from audits, process hazard analyses, or the maintenance of process safety critical equipment.

Weaknesses and Opportunities

When employees fell prey to victimization, it represented an apparent weakness to overcome. For example, shop floor employees at the effluent disposal section perceived that managers were there to harass them if they were not continuously working. They felt that they needed to be "on guard" during any inspections. These employees also insisted on "pre-safety on-site inspection" before any management safety inspections to avoid any confrontation with the executive management.

It was unclear if employees fully understood the current reduction in occupational safety incidents did not necessarily imply proactive measures in implementing and sustaining process safety. Plant managers had not adequately demonstrated to employees that reduction in occupational safety incidents should not encourage complacency in sustaining process safety. The occurrence of process safety incidents may be less frequent; however, the severity or consequence can be catastrophic.

Plant managers needed to demonstrate more process safety leadership behaviors regarding the continuous reduction of process safety risk as an ongoing improvement strategy. There seemed to be a culture of mediocrity where lowering process safety risks was balanced against 'cost versus benefits,' and there was no bias for continual risk reduction. Plant managers needed to emphasize exceeding minimum process safety and legislative requirements instead of creating a compliance-based culture.

Shop floor employees did not understand the on-site process safety hazards or risks. They had a limited understanding of the impacts of personal behavior on the controls of significant hazards and could not distinguish between slips, violations, lapses, or mistakes when executing process task operations. They were highly reliant on checklists to do their jobs and believed these checklists should be updated only after an incident when it was too late. Employees understood the slogan "Stop It for Safety," however, steam utilities plant interviewees had reservations on the use of this slogan. Employees felt that only managers were allowed to make the call of stopping unsafe acts.

Although employees were trained on process safety information or sub-contractor safety management, they still did not know the process safety criteria that needed to be included in key contracts affecting availability, procurement, and installation of safety critical equipment. Employees never questioned process safety information, and there was no discernment regarding the quality of information presented to them regarding equipment data packs, design calculations, or even control panel layout best practices. Employees still need to develop a process safety leadership lens so that they can effectively articulate the process safety standard requirements described by OSHA PSM.

Threats

When employees do not trust management, it creates a blame culture. Employees at the steam utilities plant felt disempowered and perceived that they cannot stop work when it was unsafe. That occurred because only the engineering manager was allowed to make work stoppage decisions during dangerous conditions. Effluent disposal operations employees also perceived that safety management inspections were geared towards checking if employees were working and managing equipment reliability. They believe safety was not adequately emphasized. All interviewees indicated that safety specialists needed to be more visible on the plant to spot high-risk activities and to identify hazards in the work environment.

Incident Investigation and Organizational Learning

Strengths

Regardless of the safety consequence or impact, all safety incidents were investigated thoroughly with the same depth and rigor to improve incident investigation and facilitation skills. Every process safety incident was classified according to severity, and a fire, explosion, and release index was calculated and tracked. All meetings began with a safety moment, and leaders were required to talk about a specific incident at the workplace and share incident learning to stimulate interactive dialogue and debate.

Weaknesses and Opportunities

Interviewees indicated lessons learned during incidents were not regularly shared with all employees since managers from affected business units felt that the organizational reputation was at stake whenever an event occurred. They feared being subjected to unwarranted criticism and liability; therefore, these lessons were not distributed, possibly contributing to repeat incidents. Employees realized the need for organizational learning and retaining corporate memory; however, more value-driven leadership was needed at the management level to remove a culture of fear and blame.

Insufficient reporting of incidents was due to concern for the loss of reputation, blame, a punitive culture, management, or by fear based on past experience. Also, managers set unrealistic performance targets such as zero safety incidents, which discouraged effective incident reporting. Safety moment sessions held during meetings only discussed lessons or root cause analysis after the incidents occurred and not leadership behaviors or attitudes related to process safety incidents.

Threats

Interviewees reported incident investigation lessons were ineffectively communicated, and repeat incidents have occurred. Lessons from incidents and associated learning took a long time to materialize. Experience suggested the fear, blame, or punitive organizational culture did not allow for continuous improvement or the ability to exceed minimum compliance requirements.

MANAGEMENT'S RESPONSE TO HUMAN FACTORS FINDINGS

Gas-to-Liquid and Methanol Management Teams

The gas-to-liquid management team had an "us and them" attitude towards the human factors interview feedback from their plant operators. The management team struggled to accept the need for improvement and address the concerns raised by control room operators regarding:

- The difficulty of seeing the process operating parameter font color.
- The need for maintenance procedures on process safety critical equipment instead of only using checklists.
- Operational discipline effectiveness or the difficulty experienced by employees to adhere to procedures when tools need to be lifted across elevated heights.
- The need for adjustable CCTV cameras at the gas-to-liquid methanol plant to identify any process safety incident.
- The need for additional level control alarms as an extra protection layer. Employees currently use high-level alarms as operating parameters instead of complying with the latest American Petroleum Institute codes.

Gas-to-liquid plant and control room operators provided the highest human factors perception scores despite the lack of cooperation from their management teams.

Effluent Treatment and Disposal Management Team

The effluent treatment and disposal management team had a laissez-faire attitude toward human factors feedback. There was no accountability when responding to human factors behavioral issues since team members preferred addressing recommendations arising out of the interviews and assessment instead of discussing them with their front-line employees. There seemed to be resistance toward unifying the front-line employees and the management team with one common safety goal. Operator employees often feared management team members and preferred not to have frank and honest discussions with them regarding human factors or improving safety performance effectiveness.

Steam Utilities Management Team

The steam utilities management team welcomed the findings and recommendations of the human factors assessment. They were already aware of some of the concerns previously raised by their employees.

The plant manager provided a commitment to addressing recommendations in the assessment while control room operators and front-line staff fully supported the plant manager and his conviction about sustaining process safety. It is interesting to note the steam utilities human factors perception survey exhibited the lowest score in each dimension compared to the other plants, yet the management team and employees performed as a single functional team to address their process safety shortcomings. The steam utilities management team and their employees were exposed to a "Safe Visible Felt Values-Driven Leadership" program for the three years, which allowed their staff and management to function as an effective safety team.

Risks Arising from Human Factors Assessment

The SWOT analysis identified several process safety human factors risks that can be prioritized together with preventative and corrective controls. These risks include those associated with maintenance systems, safety communication, equipment labeling, the need for operating procedures and employee competence, and process alarms and safety systems. The process safety human factors risks are consolidated below for each dimension and for all plants based on the interview feedback and survey results. Table 3.1 shows the medium-high level risks. The human factors Bow Tie or risk molecule is found in Appendix C.

Tables 1C to 7C in Appendix C include human factors risk, consequences, the inherent risk measure, preventative and corrective controls, and residual risk. The risk is assessed according to Probability (P), measured on the X-Axis, and the Impact (I) is measured on the Y-axis, as seen from the organizational Risk Matrix in Appendix C. Inherent risk excludes any preventative and corrective controls, whereas residual risk includes control measures. Residual risk is always less than the inherent risk. The cost of implementing preventive and corrective human factors risk controls versus the benefits derived was based on risk tolerability and "as low as reasonably practicable" ALARP decision making framework. Preventive controls typically reduce the chances of the risk occurring, whereas corrective controls are used to lower the impact of the risk if it does happen. The risk priority based on the level of inherent risk can be used to justify resources and CAPEX budgeting in order of increased importance. The Bow Tie methodology can be referenced in the guidelines described by CCPS and Energy Institute (2018).

Table 3.1. Summary of human factors risks

Human Factors Risk description	Proba-bility	Im-pact	Inher-ent Risk Rating
Maintenance Management	P5	I5	Level 2
Process Alarms and Safety Systems	P3	I6	**Level 3**
Safety Communication (equipment labeling and 'touch and tag' inconsistency)	P4	I7	**Level 1**
Staffing Levels and Shift Work	P5	I4	**Level 3**
Standard Operating Procedures	P5	I7	Level 1
Safety Communication (lack of hardware)	P4	I5	**Level 3**
Safety Culture and Incident Reporting	P4	I7	**Level 1**

Table 3.1 shows there are three Level 1 inherent risks: addressing safety culture and effective incident reporting together with distributing the associated incident learning, the inadequate compilation of SOPs, employee competence assurance, and inconsistent equipment labeling for pipe installations across sites. The table also shows that there is no existing process methodology applicable to the 'touch and tagging' of equipment used in the permit to work process, which is inconsistent through the various plants.

Ineffective maintenance management applicable to process safety critical equipment (PSCE) is identified as a Level 2 risk. Plant management needs to develop a robust strategy to address the PSCE management and equipment identification, as well as how PSCE will be prioritized among other equipment. Staffing levels, shift work, and safety communication are a Level 3 risk and can be managed at the plant level without executive or corporate decision-making. The recommendations for these risks are explored in Appendix C.

The human factors assessment and survey, including the SWOT analysis, used to identify common themes derived from process safety audits, team leadership behaviors, and the number and severity of process safety incidents, are described in Chapter 4. Process safety implementation and sustainability, together with the level of progress made, classification of process safety incidents, and level of reporting, are evaluated with the human factors assessment to identify similarities and differences. Team leadership behaviors underpinning human factors related behaviors that encourage or undermine process safety sustainability progress are discussed in the next chapter.

CHAPTER 4 ANALYSIS OF PROCESS SAFETY AUDIT RESULTS, FER-SI AND LEADERSHIP RESULTS

Process Safety Audit

Overview of Process Safety Audit Framework

Process safety audits are conducted on three levels: first party, second party, and third-party audits. The audit methodology described in Table 4.1 identifies advantages and disadvantages for each audit and the associated responsibilities.

Table 4.1. Process safety audit levels

Audit Level	Area of Responsibility	Advantages and Disadvantages
First-Party Audit	Plant manager and employees conduct their own audit in their plant.	Much cheaper and effective option Plant manager and the team cannot identify their process safety blind spots and over-rate their maturity in process safety due to image and reputation factors.
Second-Party Audit	Independent process safety engineers are responsible for conducting the audit.	Objective findings are obtained. Plant managers can compare themselves to their peers regarding process safety maturity.

		Audits are in-depth and test on-site and management safety system effectiveness. Process safety engineers are unable to easily find externally based best practices as they all work for the same organization.
Third-Party Audit	External and internationally recognized companies conduct process safety audits.	International best practices are identified. Global best practices can be identified, especially for safeguarding scenario identification, Bow Tie methodology, Risk Revalidation studies, and smart PSM processes. Audit scope focuses on safety management system effectiveness at a superficial level due to time pressure. These audits are expensive and time-consuming to arrange

Note: In certain instances, organizations can arrange to have a blend of second and third-party audits, where process safety engineers can aid third party auditors regarding areas of concern.

The audits discussed for gas-to-liquid, steam utilities, and effluent disposal plants were based on second-party audits and an external third party audit utilizing the blend philosophy described above. Only a first-party audit was conducted at the ammonia plant.

Process safety audit scoring guidelines are described in Table 4.2 and are used as a comparative measure to assess process safety implementation, culture, and maturity while excluding sustainability. The scores are calculated as a percentage for each process safety standard described in Appendix A.

Table 4.2. Process safety auditing guidelines

Score	Status	Guidance notes:
0	Not implemented	Extensive failures to meet the requirement were noted. The audited entity either did not implement the requirement or failed to maintain the systems that deliver compliance and thus is not in compliance with the requirement. It is possible there is a lack of awareness, understanding, or knowledge of the requirement among key personnel. There is no plan either to implement the requirement or to correct the lack of compliance if a current system falls into disuse.
1	Not implemented, but a Gap Closure Plan exists	Extensive failures to meet the requirement were noted. The audited entity has failed to maintain the systems that deliver compliance and thus is not in compliance with the requirement. There is either a plan to implement the requirement or a corrective action to correct the lack of compliance if a current system falls into disuse.
2	Partial compliance	Significant information is available, but not enough to fully comply with the standard. Incomplete documentation constitutes only partial compliance. Lack of verbal-only, formal methods of handling changes to process information also constitutes partial compliance.
3	Full compliance	The available information is sufficient to verify full compliance to the standard. The documentation is complete and easily accessible.
NA	Not applicable	The requirement is not applicable to the entity being audited or assessed. A score

		marked as NA is not considered when calculating the percentage compliance for that standard.

Source: Data from PSM OSHA, 2007.

Human factors-related process safety elements, such as employee participation, employee training, process hazard analysis (risk assessments), MOC, process safety information, and maintenance management, are assessed for common findings. Audit scores are compared for each department.

Common Findings from Process Safety Audits

The common audit findings are derived from three plants: the steam utilities, effluent and disposal, and gas-to-liquids plants based on executed second and third-party audits and first-party audit findings conducted at the ammonia plant.

a. Employee Participation

Communicating PHA risks to employees and assessing their understanding of these risks is considered a one-time-only event and is not performed continuously or used to refresh operator employee knowledge. Management of PHA risks is communicated to management teams by the employees using email. Employees, contractors, and trade union representatives may fear or not trust management regarding the communication of risks. No evidence is shown regarding whether employees understand the key loss of containment risks.

A "waiting" culture is observed at these plants since employees believe that process safety risks can only be communicated to a broader audience when PHA reports and action items arising from the PHA are approved by the plant manager, instead of immediately identifying loss of containment and proactively managing these risks.

A best practice was identified at the effluent and disposal treatment regarding the use of suggestion boxes to effectively address all safety risks.

b. Standard Operating Procedures (SOPs)

SOPs are not compiled in the correct format, and there is no logical flow of steps associated with safety critical tasks. SOPs are unavailable in the control rooms. Critical task observations are excluded from the safety management system, and it is unclear when the SOPs or checklists were updated. Most plants use outdated SOPs, which may increase operational and safety risks. Employees perceive safety critical task observations as a paper-based exercise. Typical start-up, steady-state and emergency shut-down procedures, safe operating limits, cause-consequence matrix, and troubleshooting guides are not well documented. Employees also can deviate from procedures due to lack of supervision, apathy, or the need to take short cuts due to time pressures. Informal systems are in place for the notification of expired procedures, updating, distribution, and access to procedures.

c. Training

A training matrix identifying the required competencies, knowledge, behaviors, and skills needed for operational and safety governance roles is not compiled for each job level. Chemical health hazards training, critical task training, and testing for understanding are not conducted thoroughly.

There is no formal system to track refresher training, especially when operators are transferred to different plants or when they return after a prolonged leave of absence. Certain plants require supervisors to conduct on-the-job training without formal assessment qualification, but there is no formally structured theoretical and practical competency assessment framework.

d. Process Safety Information and Hazard Analysis

The gas-to-liquid plant has an effective document management system (DMS) for Process Safety Information (PSI) and relates drawing numbers to interlocking systems. Unfortunately, this practice is not shared with other plants. Other plants have inadequate integration with other management systems, information accessibility, and revision control. Often new employees do not know the location of electronic documentation on existing DMS platforms, or they use outdated SOPs and PSI.

Updated drawings and the supporting PSI examples, such as block flow diagrams, energy balances, and process diagrams, are excluded from PHA reports. Safe operating limits (SOL) and consequences of fires, explosions, or releases if the SOL are violated, do not exist in certain plants. This includes a relief system design basis, the provision of inadvertent chemical reactions, and the safe disposal for out-of-specification products.

Baseline PHA risk assessments are completed; however, it is unclear if these risks are fully minimized, understood, and available to all employees and contractors. Consequently, previous fires occurred due to inadvertent chemical reactions and employees introducing out-of-specification products in the wrong drain.

Proactive measures are used to manage PHA risks at the gas-to-liquid plant as all disciplines fully participate, excluding the maintenance manager. Maintenance-related risks that cause fires, explosions, and toxic releases can be excluded from the PHA at the gas-to-liquid plant. The facility site layout studies and risks generated by other plants that would impact effluent and disposal treatment, steam utilities, gas-to-liquid, and ammonia plants also are not understood completely. Experience indicates there is uncertainty from plant managers regarding commitment towards implementing preventative and corrective recommendations to minimize PHA risks.

e. Management of Change (MOC) and Pre-Start-Up Safety Review

MOC usually is integrated with the Pre-Start-Up Safety Review (PSSR). Any change applicable to organizational restructuring, leadership change, legal change, retrenchments, hardware, and software modifications is managed by using a combination of safety, organizational and legal change checklists that are fully integrated within the MOC and PSSR framework. Organizational restructuring, retrenchments, and legal changes were introduced in the MOC framework with the intent of identifying and managing risks and creating a smooth transition with minimum business interruption. The PSSR provides plants with the necessary checklists to encourage facilitated dialogue discussions with various disciplines. Behavioral safety and leadership psychologists, including technical safety engineers, are used to promote and successfully manage organizational and legal changes.

It is unclear how information is edited, tracked, changed, introduced, or excluded from the MOC process. Organizational and legal changes were not fully initiated with the new MOC-PSSR framework during the recent restructuring.

The number of activities that need to be executed for each change was tracked ineffectively in certain plants, and critical stakeholders did not attend MOC-PSSR facilitated discussions or risk assessments. The overall business risk in affected plants may increase since changes are not tracked for total closure, and employees may not be fully trained or informed of the changes.

Employees are unclear about their roles and responsibilities when executing PSSR and priority recommendations are not communicated in writing. A silo mentality is visible since external specialists, including safety engineers, are not used, and other disciplines are not recognized at the commencement of the PSSR. Communication at the effluent and disposal plant also is ineffective. In one instance, employees started the plant operation when the legally accountable manager rejected authorization to begin. Plant inspections were not conducted before PSSR at the effluent and disposal treatment plant, and employees relied on memory to make knowledge-based decisions when identifying and managing risks prior to plant start-up.

f. Maintenance Integrity

Ineffective project plans noticed in all plants were used to manage maintenance integrity. These plans were compiled at a strategic level, and no identification of process safety critical equipment was conducted based on OSHA PSM requirements. A best practice was established at the gas-to-liquid plant where the equipment reliability IT platform supporting asset integrity management is used to manage maintenance and reliability of process safety critical equipment.

Encounters with plant managers indicate they are struggling to implement and sustain maintenance integrity due to inadequate resources. There is no long-term strategy on how maintenance integrity on process safety critical equipment can be integrated with other management systems, including operations excellence, reliability engineering, cost leadership, and procurement. Maintenance integrity work planning, testing and inspections, preventative maintenance, quality assurance, and training on maintenance of PSCE are not yet fully established.

g. Emergency Planning and Response

The regular testing of emergency response plans (ERP) in both physical and table-top exercises to monitor and improve employee behavior during emergencies needs to be planned and executed. The lessons from these emergency exercises are not shared with other plants. Safety critical communication during emergencies is not addressed in the ERP procedures. Because employees and contractors are not provided with adequate training, they do not know how to contact emergency services or where to gather in the event of an emergency. There also is no indication that the plants communicate their risks to close-by plants or that they address risks arising from downstream inadvertent chemical reactions that may cause a fire, explosion, or release to affected customers and suppliers.

Emergency response equipment sufficiency and location, including effective labeling, are not found in all plants. Maintenance inspection of emergency response equipment is not identified clearly or documented in inspection logbooks. There is no inventory list of emergency equipment, such as fire hoses and clamps, and some areas do not have adequate fire hoses due to theft, thus increasing process safety risks in the event of a fire.

h. Permit to Work

The permit to work audit indicates that clear roles and responsibilities and effective training were not provided to employees. This includes foremen who approve commencement of work even though they disagree with the method of energy isolation bypass. Another example is the person accountable for electrical substations does not authorize the issuance of permits for work to be done and instead insists production personnel approve work permits. Usually, the owner of the process is required to issue a permit for work to commence. Unauthorized persons, such as cleaners, have unrestricted access to substations, which could increase plant risks.

The permit is exploited and used for multiple tasks in different areas to save time, and expired permits are not renewed within 24 hours. Plant management does not authorize seven-day permits, which deviates from the prescribed permit to work procedure.

Prescribing the correct personal protective clothing, using valid permits, indicating the emergency gathering and evacuation points, monitoring of gas testing at regular frequency, calibration of gas equipment and positively identifying equipment (the 'touch and tag' principle) on the field and communicated to contractors or employees prior to any safe work execution are not conducted adequately. These occurrences undermine operational discipline and encourage a silo mentality.

In addition, the number of safety standbys is inadequate at prescribed competency levels when work, such as welding, is executed or when one standby is used for several tasks. This practice can create process safety risks since there is no dedicated personnel to watch over the execution of the work. Also, ineffective safe making and the demarcation of the work area are due to the lack of operational discipline and a shortage of a skilled safety workforce.

i. Contractor Management

The role of contractor management is not addressed in the organization at the operational site level, and plants have difficulty when escalating contractor management concerns to the corporate level. There is no formalized process for plants to address contractor safety concerns or to communicate nearby plant risks to contractor employees. The safety information given to contractors' employees is outdated, and there is no effective auditing of these safety files for content, addressing task and plant-specific risks, and for the competency of contractor employees performing any task.

A best group practice was identified regarding the risk ranking and auditing of all contractors for safety maturity and ethics.

Summary of Common Audit Findings

The audit findings indicate employees and contractors do not participate in identifying, understanding, and managing process safety risks. Pockets of excellence are identified in different plants. For example, the effluent and disposal plant uses suggestion boxes, and the gas-to-liquid plant has an effective document management system to change or update information and equipment reliability systems. However, these methods are not shared with the remaining plants, undermining the potential for organizational learning. Risks from adjacent plants and contractor employees' safety also are not addressed effectively. Instead, each department manages its own risks and does not consider the consequences or risks imposed upon or by other plants. There is no effective communication of risks or lessons learned to other plants, nor is there any information sharing with affected stakeholders about emergency exercises or previous incidents. Contractors are treated differently from the organizations' employees, and no provision is made to address contractors' employee competency, safety, and risk.

The permit to work process is audited weekly by departmental employees, and second party audits are conducted biannually. The permit audits are not used as a yardstick to enforce operational discipline or address continuous improvement, and operational violations are repeated across the plants. Employees feel burdened by the lengthy work permit execution process. Moreover, the organization over emphasizes the permit to work process as the last layer of defense and marginalizes resources allocated towards compiling SOPs. If high quality, sufficient SOPs are assembled and include maintenance activities together with effective employee training, then the number of work permits issued would decrease, and there would not be an over-reliance on the permit to work process. Then, the SOP and training system would become an effective layer of protection in addition to the permit to work process.

The long-term maintenance integrity strategy is misaligned with the available manpower resources. There is no distinction between regular and statutory maintenance and process safety asset integrity due to the difficulty in identifying PSCE. Employees are confused and feel burdened as they perceive asset integrity, which does not contribute to the triple bottom line as an additional responsibility. The organization should consider using operating quantity thresholds of hazardous chemical substances, equipment reliability and availability, process risk, and operating parameters as a function of prioritizing PSCE.

A quantitative evaluation of the process safety audit scores for each department is assessed for implementation progress and the identification of common trends. Ineffective communication, strategy formulation, and decision-making weakened process safety implementation and sustainability due to the silo mentality and the blame and "waiting" culture behaviors, which is explored further in the Barret survey team leadership assessment for each department.

Process Safety Audit Scoring Analysis

The process safety implementation audit scores are shown in Figure 4.1 for 11 process safety standards, excluding incident investigations, compliance audits, and trade secrets. A first-party audit was conducted on the ammonia plant, and second party audits were completed in the remaining plants. Based on experience and advice given by third-party auditors, usually, a score of more than 85 percent is required for effective process safety implementation.

The ammonia plant process safety audit, which was conducted by its employees, has scores exceeding 85 percent. The scores are inflated due to employee bias and the desire to uphold their organizational reputation and image, thus compromising organizational learning and continuous improvement.

Figure 4.1. Process safety audit scores using OSHA CFR1910 framework. Data from Restricted, 2010.

Second-party audit results, conducted by independent technical safety engineers, provide objectivity. In this audit, lower scores are indicated for all process safety standards, thus supporting the need for continuous improvement and highlighting deficiencies and the need for resources. Employee participation has the lowest score. The lack of effective and consistent communication and the need for updated PSI is evident in the plants. Also, adequate standard operating procedures and effective training are required, as indicated by the human factors perception survey, feedback from interviewees, and audit scores below 85 percent.

Although employees are satisfied with the MOC and PSSR process, as noted in the process safety audit scoring and human factors assessments, the need for change management during organizational restructuring and transformation was not addressed fully in the audit. The contractor management scores are more than 60 percent for all plants; however, the safety of contractors' employees is not entirely addressed in the audit since there is no process to address contractor employees' safety concerns.

PHA risk assessment progress is observed in all plants with scores ranging from 45 to 83 percent; however, the effluent and disposal plant needs to provide a PHA plan and time completion dates for all risk assessments. The permit to work audit scores are higher than 90 percent for compliance to the work permit process and is used as a last barrier of protection in preventing an incident before any work is executed, according to the Swiss cheese model. The SOP and training scores are lower compared to the permit to work compliance process since manpower resource allocation is reduced to compile SOPs and conduct employee training. The organization can increase the safety culture maturity by institutionalizing operational discipline through compilation of SOPs and training, instead of only emphasizing permit to work compliance.

Inadequate progress was made for the implementation of maintenance integrity. Scores ranged from 52 to 90 percent due to a shortage of skilled workforce resources, uncertainty regarding the identification of process safety critical equipment, and failure to integrate the maintenance integrity roll out with reliability engineering and operations excellence.

The audit results indicate that there is a need for operational discipline, resource allocation, and utilization, requiring priority and higher executive committee management commitment to implement and sustain process safety fully.

The effectiveness of process safety implementation and sustainability progress discussed in the audit results is related to the organizational safety maturity culture and is a function of the number of incidents resulting in FER. The FER-SI and FER-SR are lagging indicators, triggering which management systems are failing and identifies trend patterns and incident root causes.

The Need for Process Safety Balanced Score and Hydrocarbon Leak Analysis

The organization did not select an approved set of leading and lagging process safety indicators in addition to hydrocarbon leak management. A popular set of scorecard indicators, shown in Figure 4.2, is based on recent trends observed in major energy companies that focus on asset integrity process safety management. The methodology used for scorecard selection is based on design, operating, and successfully sustaining process equipment integrity leadership, as shown in the asset integrity pyramid in Figure 4.3. The three process integrity dimensions (design, operate, sustain) are integrated with a human-machine interface and abnormal situation management reliant on a robust implementation of a process safety management system. Process integrity leadership behaviors are supported by a mature process safety culture composed of inter-relationship behaviors of the individual, organization, and society. Leading and lagging indicators are best selected to measure each dimension of the asset integrity excellence pyramid.

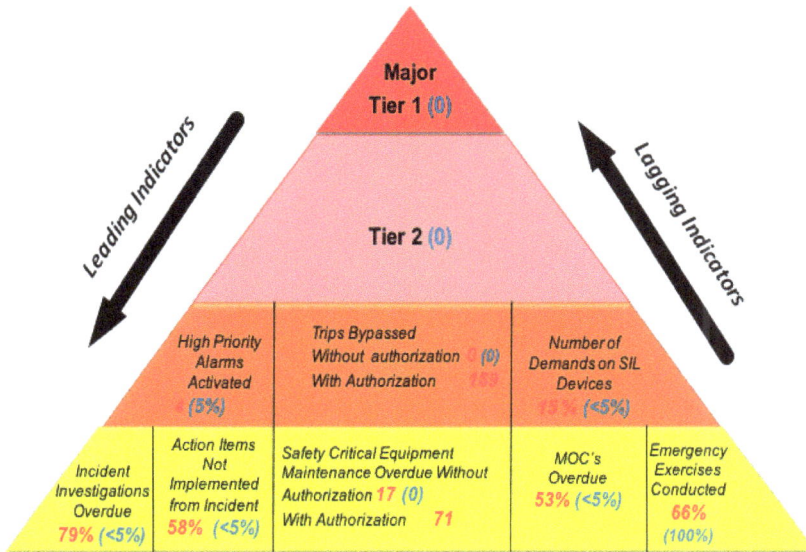

Figure 4.2: Popular leading and lagging indicators (excluding leadership behaviors).

Leading and lagging indicators are used to monitor the process safety balance scorecard. Some of the popular indicators used for performance monitoring are described in Figure 4.2. A leading indicator measures a future event and emphasizes implementing activities to prevent fires, explosions, or toxic releases (Ergo-plus, 2017). Lagging indicators measure failure while leading indicators measure performance.

Daniels (2017) states that leading indicators should:

- Demonstrate incremental process safety performance improvement
- Measure positive attributes, such as what stakeholders are implementing versus failing to do
- Encourage regular feedback to all stakeholders using Process Safety Steering Committee Meetings
- Be measurable and create a bias for action
- Be predictive based on year-to-date or month-to-date data trends

- Encourage constructive problem solving around process safety while setting key milestones
- Assist with developing specific, measurable, achievable, realistic, process safety action plans with defined completion timelines.

There is no "one size fits all" measure for safety; however, following these criteria assists in tracking impactful, leading indicators. Lagging indicators measure process safety incident statistics and, in most cases, satisfy progress made toward compliance requirements.

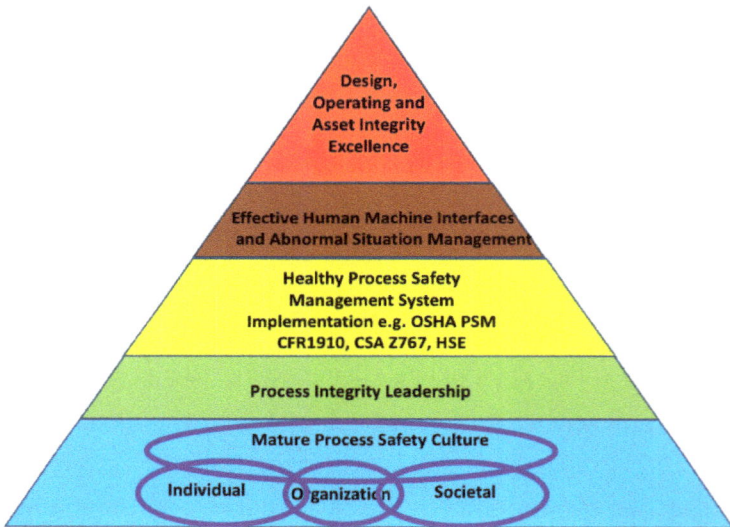

Figure 4.3. Asset integrity excellence pyramid.

The organizational FER-SI toolkit, developed by CCPS (2009) and customized by the organization, is a lagging indicator of performance and measures process safety incident severity and classification. A procedure was compiled describing the use of the tool to analyze trends, manage process safety incidents and provide management with an accurate indication of the seriousness and magnitude of process safety incidents occurring in their areas of responsibility.

The FER-SI is a cumulative severity weighting for a single process safety incident. It considers all the criteria that could make the incident more severe and adds them up in a weighted fashion to obtain a severity score for that incident. These include, among others, the actual and potential consequences of the incident, the hazardous nature of the chemicals involved, and whether protective controls failed.

The FER-SR is the sum of all the calculated FER-SI for all incidents in a month and then is normalized using the hours worked. The hours worked is an estimate of the size of the operation and compensates for changes in the size of the operation. Consideration of the criteria used in calculating FER-SI already conveys an indication of what went wrong to cause the incident and supports the incident investigation process. The principle of the FER-SI is that a severity score is assigned to an incident based on select criteria. An incident is considered more severe under the following conditions:

- A larger quantity of a more hazardous chemical is released
- The actual consequences are more severe in terms of injuries, direct financial loss, the environment, the community, the company image and the time taken to stabilize the incident
- The potential consequences could have been more severe than they were
- Preventive controls in plant design and maintenance failed to prevent the incident
- Management controls, such as procedures and training, was unable to prevent the incident
- Mitigating equipment included in the design of the plant was ineffective to control the fire or release

Table 4.3 describes the criteria together with the relative weight scores. The maximum total score is calculated as a percentage out of a total score of 740.

Table 4.3. FER-SI criteria and relative weighting

Criteria
The actual quantity of the release and hazardous nature of the chemicals involved (150)
Actual injuries (100) Direct financial loss (100) Environment, community and company image (100) Time taken to stabilize the emergency (50)
The potential release quantity and hazardous nature of the chemicals involved (20) Number of people that potentially could have been injured (20)
Mechanical integrity maintenance strategies failed (50) Design preventative controls failed (50)
Management controls failed (50)
Design mitigating features failed (50)

Source: Data from CCPS, 2009a.

The monthly FER-SI is calculated by adding up the FER-SI severity index scores for all the incidents occurring in a month. To calculate the FER-SR, the monthly FER-SI score is normalized by including the hours worked by employees. This calculation excludes contractors to allow for the automatic correction for the growth of a business as the number of hours covers the size of the operation.

FER-SR = (FER-SI sum total for all incidents for the month) X 200,000

Actual hours worked

*200,000 is a recognized constant in the energy sector.

Also, process safety incidents are classified according to FER-SI and are shown in Table 4.4. The provision for any employees or contractors incurring a lost workday case is excluded.

Table 4.4. Process safety incident classification

Incident classification	FER severity index (FER-SI)
Major	45 and higher
Significant	35 and higher, but <45
Moderate	20 and higher, but <35
Minor	8 and higher, but <20
Insignificant	<8

Source: Data from CCPS, 2009a.

Analysis and Discussion of Fire Explosion Release Severity Rate

Figure 4.4 shows the FER hydrocarbon leak severity rate for the steam utilities plant. Only two incidents were recorded from July 2009 to September 2012 due to oil spillage and the resulting fire, with minor and moderate incident classifications. Figure 4.5 shows the leak severity of the ammonia plant, resulting in three minor incidents due to ammonia and natural gas releases. Effective reporting after the first quarter of 2012 was a result of the new organizational structure and merger of the ammonia plant with the rest of the business. The new plant managers provided effective leadership and insisted that all incidents were recorded for business scorecard performance and continuous improvement.

Inadequate incident reporting not resulting in fatalities and recorded during July 2009 through early 2012 was due to the punitive and blaming work culture. This is because plant managers included FER statistics in employee performance contracts, and employees felt prejudiced if the FER trends began to increase as their performance ratings would be undermined.

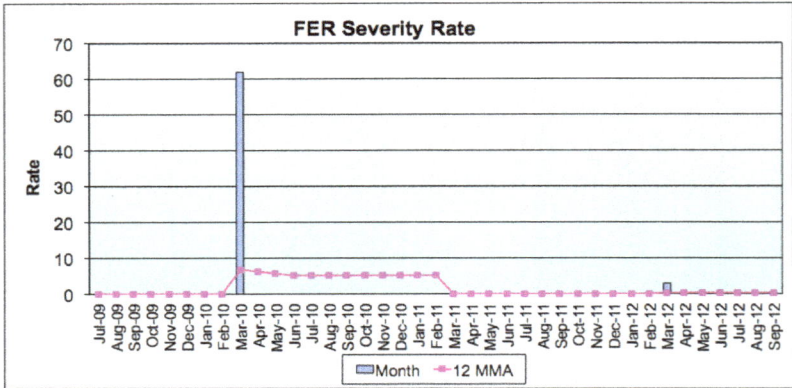

Figure 4.4. FER severity sate for the steam utilities plant. Data from Restricted, 2012b.

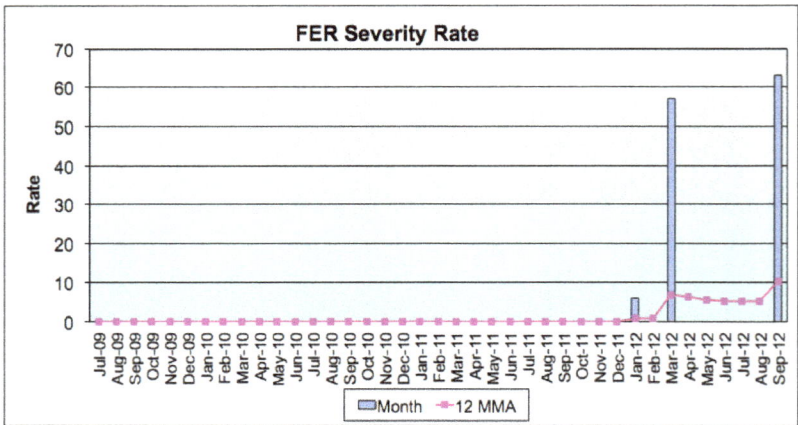

Figure 4.5. FER severity rate for the ammonia plant. Data from Restricted, 2012b.

Figure 4.6 shows the FER-SR for the gas-to-liquid plant. Effective reporting resulted in the department identifying root causes and deviances in the process safety management system. Significant and moderate incidents occurred in October 2009, while the rest were minor incidents occurring from November 2009 to July 2012. A downward trend is noticed regarding the severity of the incidents and the 12-month moving average (MMA). Most of the root causes associated with these incidents were the result of not following the MOC process or not adhering to equipment maintenance frequencies. Employees do not fear reporting incidents when the number and severity of process safety incidents are excluded in performance contracting and when the plant management continuously achieves improvement on their process safety performance scorecard.

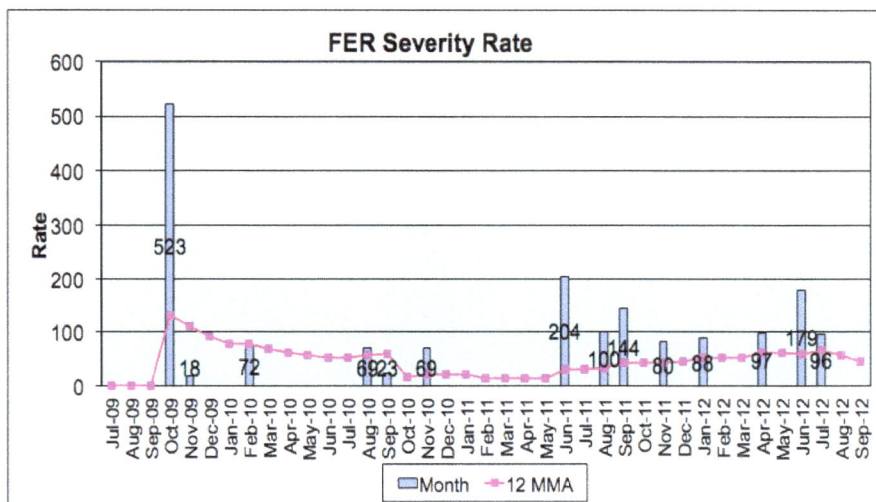

Figure 4.6. FER severity rate for the gas-to-liquid plant. Data from Restricted, 2012b.

Figure 4.7 shows the FER severity rate for the effluent and disposal plant. Adequate reporting did not start until October 2010, which may have been a result of the change in plant management. The new plant manager and management team are committed to process safety. Operational employees, however, may feel reluctant to report minor process safety incidents due to prejudice if their performance ratings may be undermined. An upward trend is seen in the 12 MMA from April 2012 to July 2012, resulting in minor incidents, while the October 2010 incidents resulted in significant and minor incidents. Ineffective reporting is noticed from October 2010 to March 2012, since minor incidents may not be easily detected.

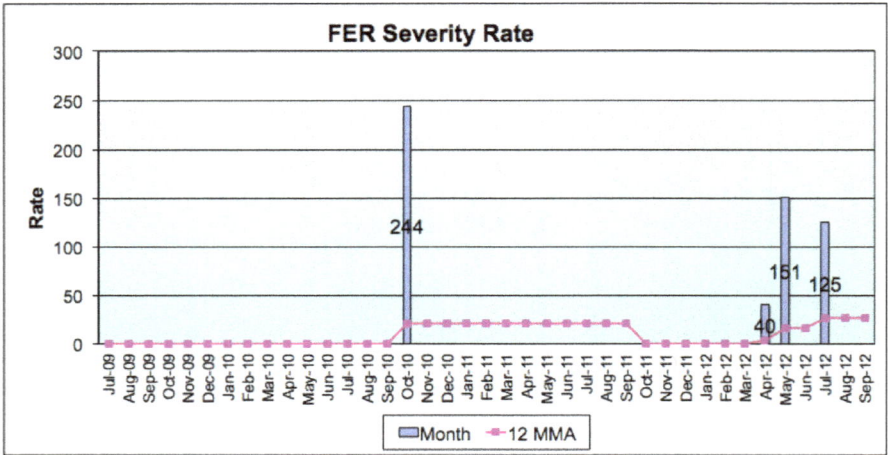

Figure 4.7. FER severity rate for the effluent and disposal plant. Data from Restricted, 2012b.

Most of the root causes for process safety incidents, as shown in Figures 4.4 to 4.7, originate from a lack of operational discipline and inadequate maintenance. These incidents are repeated in all plants due to ineffective organizational learning. Root cause analysis is a subset of risk management, and the number of process safety incidents is used as a lagging indicator towards effective risk management performance. The reporting of process safety incidents is a function of process safety culture maturity. Identifying and analyzing the root cause of incidents allow for the revision of preventative and corrective controls aimed at promoting continuous improvement. Figure 4.11 shows how effective risk management, incident reporting, and effective organizational learning are integrated to increase organizational process safety maturity.

Figures 4.8 to 4.10 depict root causes associated with process safety incidents for all plants from the second quarter of 2010 until the first quarter of 2013. There is an increasing lack of operational discipline and a reduction in asset integrity root causes, mostly due to failure to maintain equipment at the correct preventative maintenance frequency schedule.

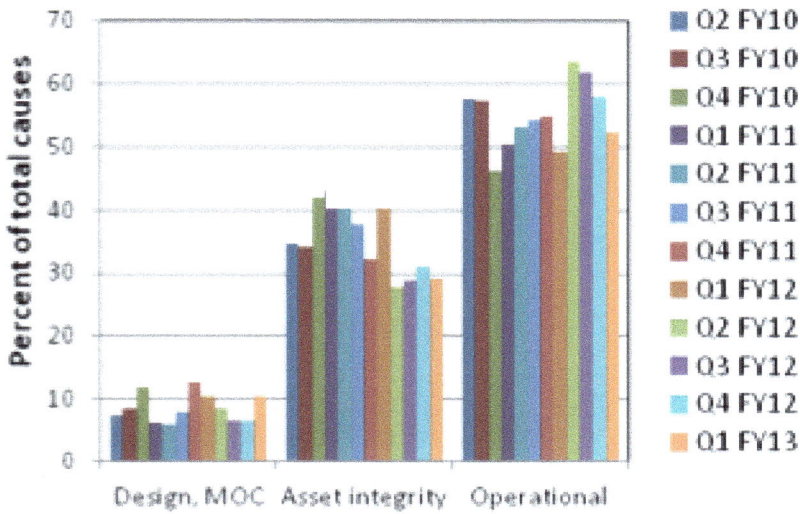

Figure 4.8. Categorization of FER causes. Data from Restricted, 2012a.

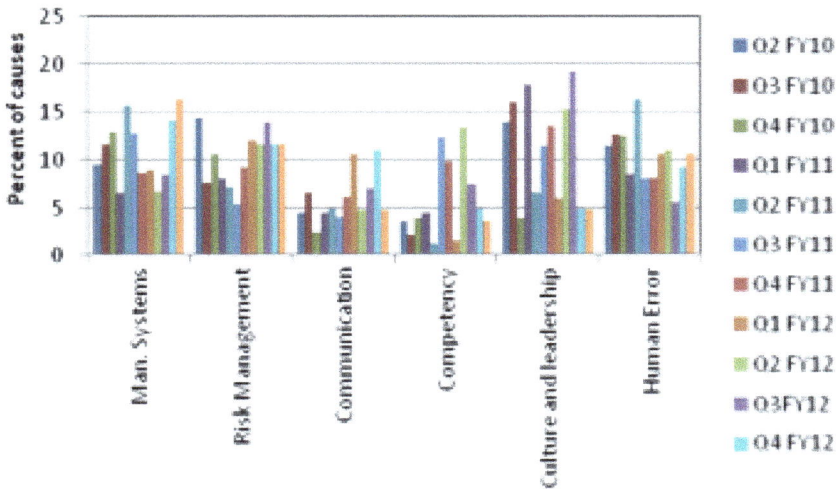

Figure 4.9. Causes of equipment failure. Data from Restricted, 2012a.

The lack of operational discipline is due to an accumulation of deficiencies in safety management systems, failure of employees and contractors to follow procedures, a lack of competence assurance and ineffective risk management. This state exists when employees do not fully comply with or initiate the change management or permit to work processes.

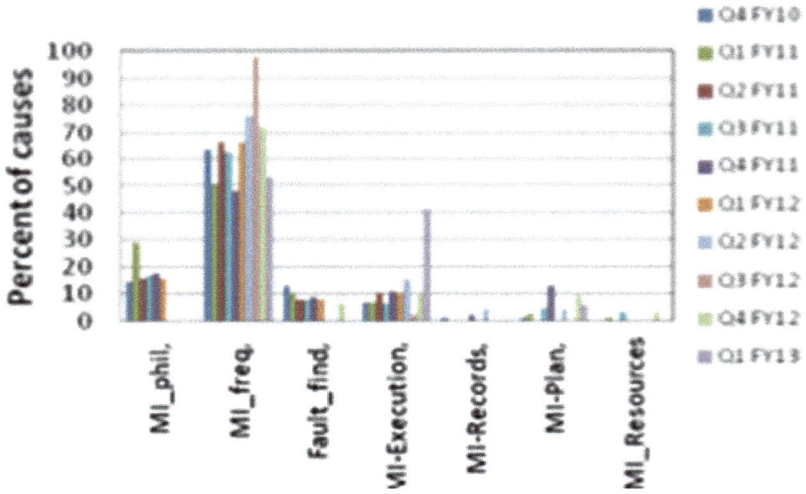

Figure 4.10. Organizational and human factors contributions to process safety incidents. Data from Restricted, 2012a.

Figure 4.10 indicates safety culture and leadership (not supported by the correct behaviors as indicated from the Barret survey) and human error because of non-compliance with maintenance-related operating procedures are the two primary drivers contributing to process safety incidents. These findings, related to ineffective supervision and procedural compliance, also are identified from the human factors survey and SWOT analysis described in Chapter 3.

The change in leadership at the executive committee level during mid-2011 and institutionalizing new corporate leadership values at the beginning of 2012 may have contributed towards a decline in process safety incidents due to culture and effective leadership changes. The new corporate values being institutionalized are simplicity without bureaucracy, clear and consistent communication, accepting accountability, removing silos, and accepting that employees belong to "one organization." Leadership development and employee accountability, together with the focus on continuous improvement, are critical success factors for accelerating process safety culture and maturity.

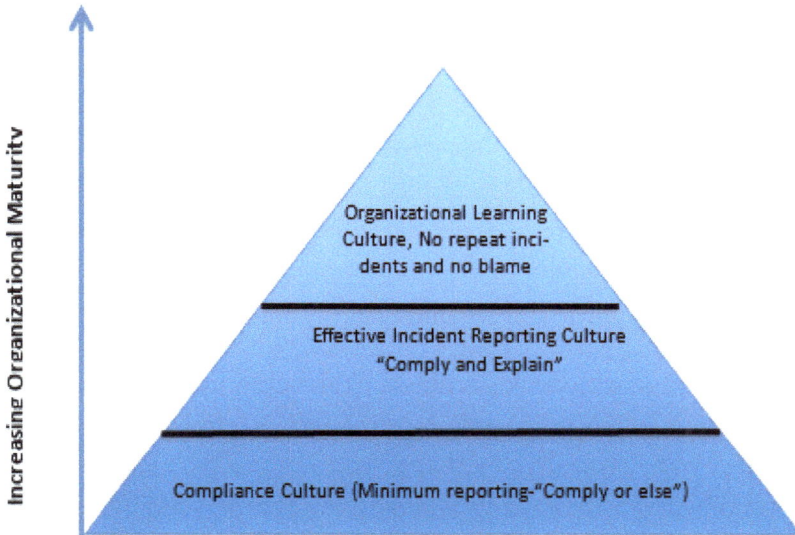

Figure 4.11. Organizational learning pyramid.

Figure 4.11 suggests that minimum incident reporting promotes a compliant or punitive work culture and leads to ineffective organizational root cause analysis. This is because process safety risks may be higher than anticipated, and no provision is made to address the required preventative and corrective controls. Effective incident reporting encourages correct and regular root cause identification, continuous improvement, and moving beyond a compliance culture. Employees explain the reasons for their mistakes and provide management with solutions to ensure that similar incidents are not repeated in all plants.

Lessons learned from safety-related incidents help the organization to prevent the repetition of these incidents in the future and allow the organization to entrench an organizational learning/knowledge management culture. Plant management teams and employees should have no fear of blame and reputation damage when sharing lessons learned from their own mistakes with other plants and business units. Leadership transformation must begin at the plant management level and in the operations environment to encourage effective knowledge sharing, incident reporting, and root cause identification.

Employee attitudes and behaviors, which have undermined process safety implementation progress related to common audit findings and were the cause of repeated incidents, are further investigated by identifying departmental team leadership deficiencies using the Barret survey results discussed in the next section.

Analysis of Barret Leadership Survey

There is a clear need for leadership development and organizational transformation, in addition to monitoring and assessing underlying behaviors that undermine rather than encourage process safety sustainability. They are discussed for each plant using the Barret Leadership Survey.

Overview of the Barret Leadership Survey

The Barret leadership survey was conducted by an independent organization and used by the four plants to assess their leadership blind spots and how the blind spots contribute to human factors concerns and safety culture maturity. Organizational effectiveness is measured using the Barret values survey Tool (Barret, 1998) and is based on seven levels of consciousness for employees and organizations (see Figure 4.12).

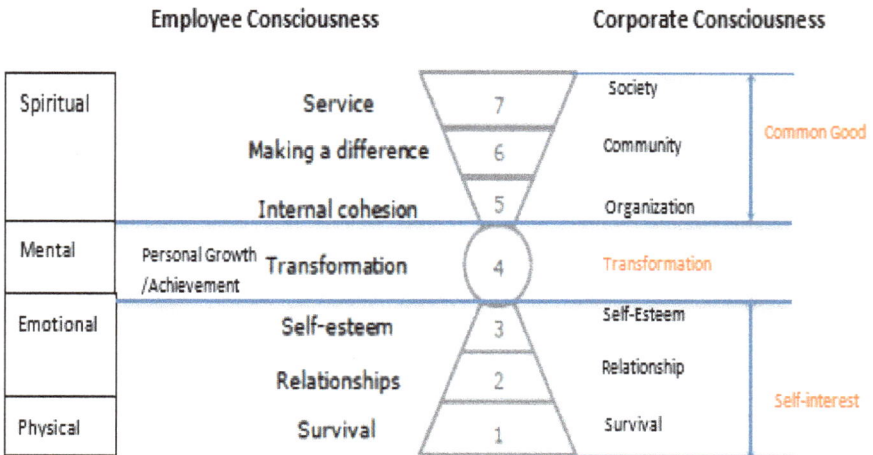

	Employee Consciousness			Corporate Consciousness	
Spiritual	Service	7		Society	Common Good
	Making a difference	6		Community	
	Internal cohesion	5		Organization	
Mental	Personal Growth /Achievement Transformation	4		Transformation	
Emotional	Self-esteem	3		Self-Esteem	
	Relationships	2		Relationship	Self-interest
Physical	Survival	1		Survival	

Figure 4.12. Seven levels of consciousness for employees and organizations. Data from Barret, 1998.

The employee personal values (I), their present organizational values (CC), and values that employees would desire in the organization (DC) are recorded and analyzed through an online survey from Barret (1998). An alignment exercise is executed to determine which values are common in the organization and values that are positive (P) or virtuous (for example, honesty, trust, and accountability) versus potentially limiting values (such as blame, revenge, and manipulation). These values are rated according to self-interest (S), common good (C), and transformation (T). Potential limiting values (L) support the ego in meeting its needs as manipulation encourages exploiting others to satisfy personal needs, and blame is used to prevent humiliation and revenge and is perceived as 'getting even.' Leadership authenticity begins to decline whenever the egoistic behaviors are misaligned with the virtuous (P) values. A cultural entropy score is computed. This score measures the amount of employee disengagement, unproductive work, or idleness in the team directly resulting from friction, frustration, or conflict in a specific plant and is calculated as a proportion of limiting values selected by employees during the survey.

Organizational leadership effectiveness of the four plants is evaluated to identify team leadership maturity level, potential limiting values, and entropy levels. The entropy levels and the identification of potentially limiting values aid in assessing the safety maturity culture and in validating the process safety audit results and feedback from the human factors interview assessments. Figures 4.11 to 4.14 show the Barret leadership values for the four plants. The results are based on a survey compiled in 2011 and analyzed by a recognized independent third party. Survey respondents included the different plant management and employee teams by department.

Gas-to-Liquid Plant Barret Survey Results

Figure 4.13 shows the positive (P) and limiting values (L) selected for individual personal values (PV), the current organizational culture (CC), and the desired cultural (DC) values. Under the Personal Values Column (PV), seven positive values (P) were selected by employees that relate to individual values (I), including honesty, commitment, integrity, safety, dedication, responsibility, and performance. However, in relationship behaviors (R), five values were selected. They included: accountability, family, fairness, respect, and trust. Employees did not select any societal values (S) as part of their personal values. Thus, the personal positive values index indicates IRS(P)=7-5-0. Similarly, employees did not select any undesired/ego-based limiting values (L) as their individual (I), relationship (R), or societal leadership values (S); thus IRS(L)=0-0-0. Employees selected values aligned to level five personal leadership behaviors, namely internal-cohesion.

The Current Culture Values (CC) indicate the values that gas-to-liquid employees selected based on positive (P) and limiting (L) behaviors regarding individual (I), relationship (R), organizational (O), and societal (S) leadership values. This translates into IROS(P)=1-2-7-0, indicating that one individual positive value (I), achievement, is related to the current department culture, while the two positive relationship values (R) of reliability and teamwork are important to team members. Seven positive organizational values (O) also are visible at the gas-to-liquid plant. They consisted of customer focus, employee safety, continuous improvement, cost reduction, productivity, quality, and performance. No positive societal leadership values are identified, indicating silo-focused or discriminatory behaviors toward non-gas-to-liquid team members. There are no limiting (L) current culture values identified; thus, IROS(L)=0-0-0-0.

The desired culture (DC) values that gas-to-liquid employees want manifested in their department related to individual values (I) are commitment, honesty, and integrity. Relationship behaviors (R) include open communication, reliability, and teamwork. Organizational (O) values are employee safety, customer focus, leadership development, and information sharing. The common values of customer focus, employee safety, continuous improvement, reliability, and teamwork are between CC and DC, while honesty, commitment, and integrity exist in the PV and DC leadership dimensions. Although gas-to-liquid team members do value safety in all three leadership dimensions, the pressure to perform and achieve results while reducing costs is persuading gas-to-liquid team members to compromise on integrity, honesty, and open communication.

The bottom portion of Figure 4.13 indicates the current organizational culture has an entropy level of 13 percent. According to Barret (1998), this score needs to be reduced to less than 10 percent for sustainable business. The main leadership blind spots preventing the gas-to-liquid team from reaching a higher safety and maturity culture (see Table 4.5) are image, arrogance, power (Level 3), and blame (Level 2). These behaviors are due to the need for individual achievement and blind ambition, as well as reluctance to accept accountability. Image behaviors were evident during the human factors interview assessment pertaining to safety culture and management visibility. Employees tried to create a positive image of their managers' safety leadership styles and did not reflect deeper. Blaming of employees by management due to fear, loss of reputation, and arrogance also surfaced when the findings and recommendations of the human factors assessment were provided to the gas-to-liquid management team. The maintenance and control instrumentation managers displayed arrogance by continually questioning the integrity of the assessor and did not believe the findings regarding ergonomics and employee difficulties when executing maintenance or process control operations.

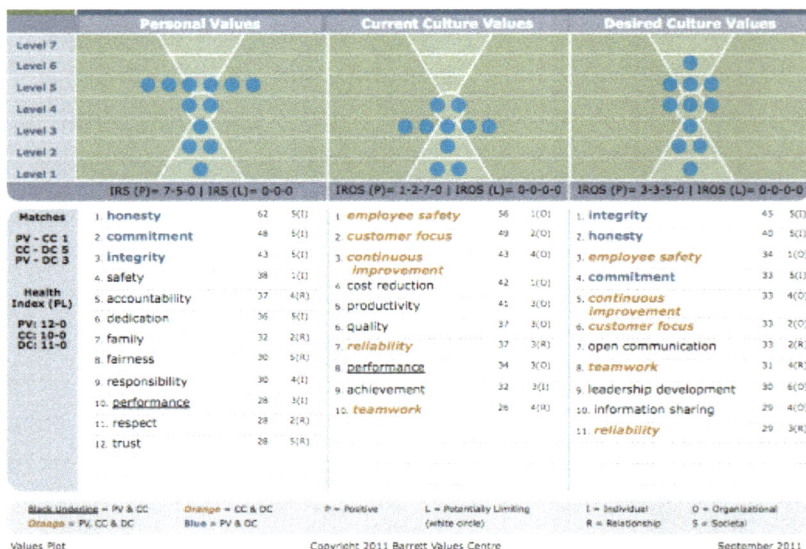

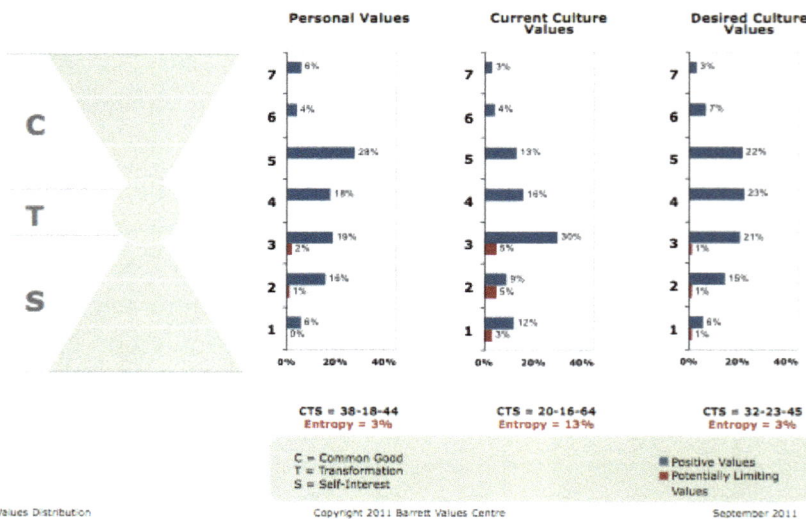

Figure 4.13. Team leadership values results for the gas-to-liquid plant. Data from Restricted, 2011.

Steam Utilities Plant Barret Survey Results

Figure 4.14 shows the team leadership values results for the steam utilities plant. Personal values IRS(P)=5-8-0 is calculated, and there are mostly level 5 and level 2 leadership values that are visible, indicating that the steam utilities team is focused on internal cohesion and employees value their relationships with others. Current culture values are IROS(P)=1-0-7-0 and IROS(L)=0-1-1-0. Organizational leadership maturity emphasizes Level 3 (self-esteem) and Level 1 (survival) behaviors due to the recent organizational restructuring leading to job security uncertainties and the appointment of a first-wave leader. Employees feel devalued or demoralized with the new leader and exhibited the lowest human factors rating (see Figure 3.1). Potential limiting values based on organization (O) and relationships (R) regarding job insecurity and blame are visible in this department due to the fear of management and restructuring. No societal leadership values are present for PV, DC, and CC, thus encouraging a silo-focused mentality, which is confirmed in Table 4.5.

The common value shared throughout the steam utilities plant is commitment and safety, and customer focus is the current and desired value states. The value of accountability is visible in the personal and desired culture states. The department has a 22 percent entropy level attributed to image, arrogance (level 1), and blame (level 2), as seen in Table 4.5. The feedback of the human factors assessment was well received by management. The plant manager, however, was indecisive and unwilling to accept accountability for implementing recommendations derived from the process safety management or human factors assessments. This situation might be due to the fact the department may be either sold off or shut down in the near future. Most of the findings and recommendations regarding the need for updated operating procedures and critical task observations made earlier to management also were identified by auditors and were not given priority attention. The steam utilities plant's culture of blame, fear, and arrogance and the exclusion of accountability in the current work culture, including the lowest scoring from the human factors perception survey results discussed in Chapter 3, indicate human factors concerns and issues are given low to no priority by the management team.

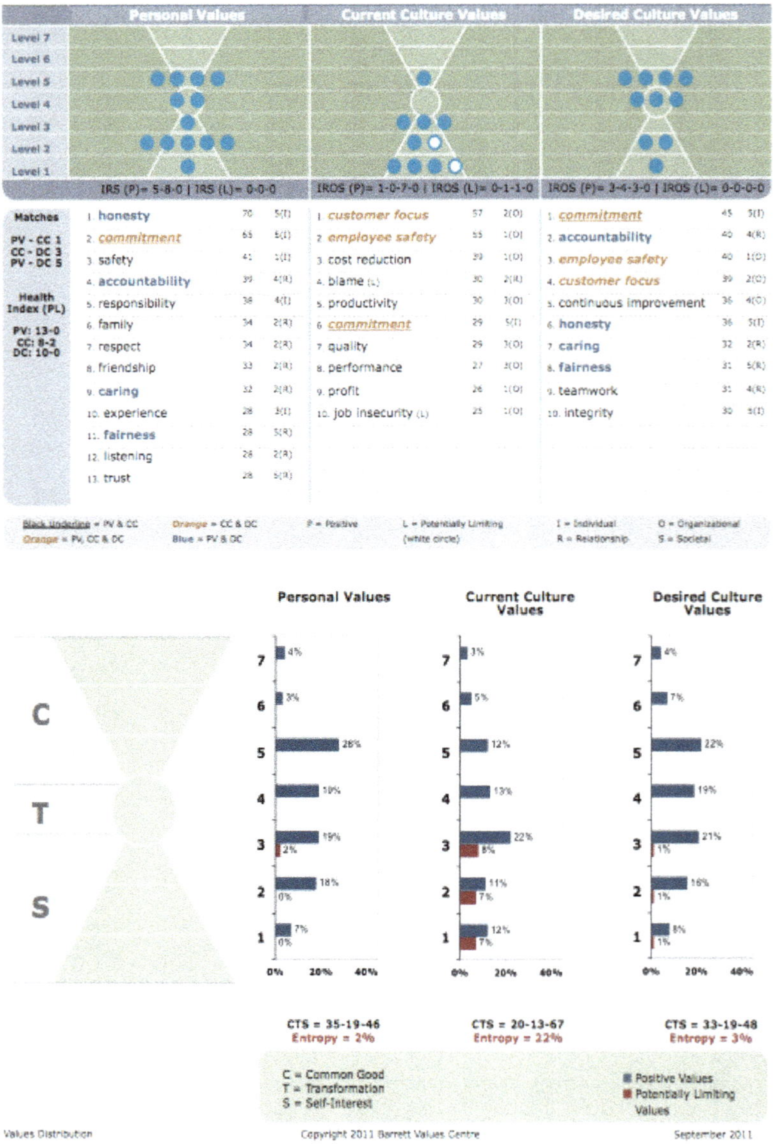

Figure 4.14. Team leadership values results for steam utilities. Data from Restricted, 2011.

Effluent and Disposal Plant Barret Survey Results

Figure 4.15 shows the Barret leadership survey results for the effluent and disposal plant. A PV of IRS(P)=6-4-0 and employee's personal values are developed from Level 5 to Level 2, indicating that internal cohesion, transformation, relationships, and survival are essential to employees. The current culture indicates IROS(P)=3-1-7-0 and IROS(L)=0-1-0-0. Most of the positive current values are related to the organization, and commitment and safety are common throughout the personal, current, and desired value dimensions, whereas accountability is excluded in the current values, and blame is identified as a potential limiting value. Thus, employees may be subjected to management by fear, which encourages a blame culture, and may be reluctant to take ownership of their roles and responsibilities. Table 4.5 shows arrogance, power, information hoarding, blame, discrimination, and empire building are the main potential limiting behaviors, which contribute to the 21 percent entropy level at the effluent and disposal plant. The feedback from the Barret survey can be aligned to the human factors assessment that indicates the plant management team only focuses on plant reliability and human productivity, and not much on the well-being of employees.

The effluent and disposal management team was reluctant to accept accountability and ownership of the human factors assessment findings and recommendations and postponed the management feedback session on several occasions. During the human factors feedback session with the management team in October 2012, the plant manager insisted he was unwilling to accept accountability for human factors related to plant maintenance issues and preferred to address the associated recommendations directly with his maintenance team. Also, he preferred to have a follow-up meeting to address the remaining human factors recommendations further. Issues of blame and abuse of power surfaced when the plant manager requested the disclosure of employees who confessed to standing on the guardrail to perform maintenance work since these actions violated work and safety procedures, despite the interview process being based on a 'no name and no blame' principle.

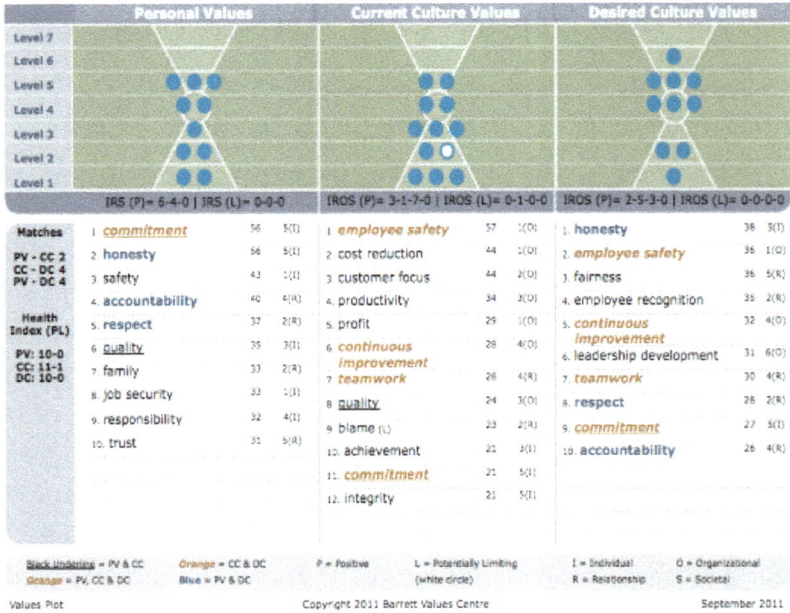

	Personal Values	Current Culture Values	Desired Culture Values
	IRS (P)= 6-4-0 | IRS (L)= 0-0-0	IROS (P)= 3-1-7-0 | IROS (L)= 0-1-0-0	IROS (P)= 2-5-3-0 | IROS (L)= 0-0-0-0

		Personal Values		Current Culture Values			Desired Culture Values	
Matches	1. commitment	56 5(I)	1. employee safety	57 1(O)	1. honesty	38 5(I)		
PV - CC 2	2. honesty	56 5(I)	2. cost reduction	44 1(O)	2. employee safety	36 1(O)		
CC - DC 4	3. safety	43 1(I)	3. customer focus	44 2(O)	3. fairness	36 5(R)		
PV - DC 4	4. accountability	40 4(R)	4. productivity	34 2(O)	4. employee recognition	35 2(R)		
Health	5. respect	37 2(R)	5. profit	29 1(O)	5. continuous improvement	32 4(O)		
Index (PL)	6. quality	35 3(I)	6. continuous improvement	28 4(O)	6. leadership development	31 6(O)		
PV: 10-0	7. family	33 2(R)	7. teamwork	26 4(R)	7. teamwork	30 4(R)		
CC: 11-1	8. job security	33 1(I)	8. quality	24 3(O)	8. respect	28 2(R)		
DC: 10-0	9. responsibility	32 4(I)	9. blame (L)	23 2(R)	9. commitment	27 5(I)		
	10. trust	31 5(R)	10. achievement	21 3(I)	10. accountability	26 4(R)		
			11. commitment	21 5(I)				
			12. integrity	21 5(I)				

Black Underline = PV & CC   Orange = CC & DC   P = Positive   L = Potentially Limiting   I = Individual   O = Organizational
Orange = PV, CC & DC   Blue = PV & DC   (white circle)   R = Relationship   S = Societal

Values Plot   Copyright 2011 Barrett Values Centre   September 2011

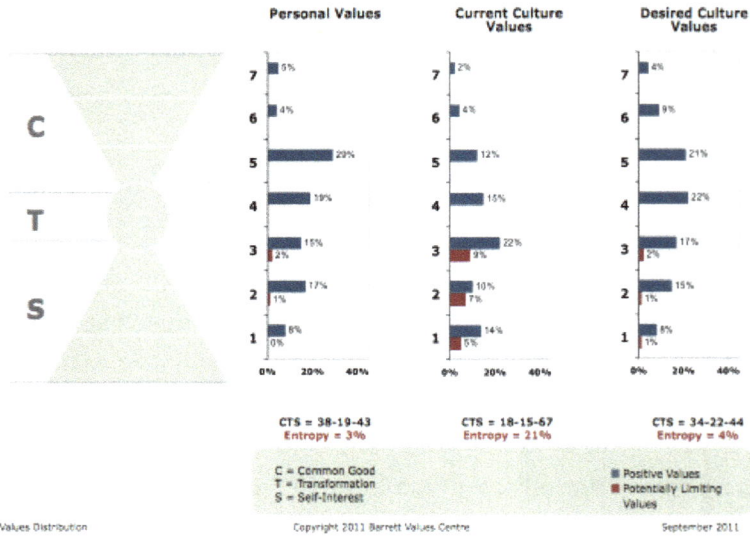

Personal Values	Current Culture Values	Desired Culture Values
CTS = 38-19-43	CTS = 18-15-67	CTS = 34-22-44
Entropy = 3%	Entropy = 21%	Entropy = 4%

C = Common Good
T = Transformation
S = Self-Interest

Positive Values
Potentially Limiting Values

Values Distribution   Copyright 2011 Barrett Values Centre   September 2011

Figure 4.15. Team leadership values results for effluent and disposal. Data from Restricted, 2011.

It is interesting to note that the thermal oxidation unit at the effluent and disposal plant experienced numerous process safety incidents in 2012 due to a shortage of skilled staff in the control room and on the field as well as inadequate operating discipline. During the incident investigations, employees readily acknowledged the blame culture by conceding that the plant manager wanted someone to blame instead of finding the root cause of the incidents.

The desired culture indicates employees require transformation and internal cohesion (Level 4 and 5 behaviors) to accelerate the organizational safety maturity. Leadership development is necessary to remove the potential limiting behaviors of discrimination and abuse of power and to encourage team building between employees and plant management.

Ammonia Plant Barret Survey Results

The value-driven leadership results for the ammonia plant are shown in Figure 4.16. Personal values IRS(P)=6-4-0 of transformation (Level 5) and relationship (Level 2) behaviors are visible, while the current culture demonstrates self-esteem (level 3) behaviors. The desired culture sought by employees shows a need for transformation (Level 4) and internal cohesion (Level 5). Safety values are noticed in all three PV, CC, and DC dimensions, and employee safety, teamwork, and continuous improvement are noticed in the CC and DC dimensions. Accountability is excluded from the current cultural values, but employees acknowledge the need to entrench this value in their desired culture.

The human factors study conducted by ERM (2007) indicated safety culture between ammonia plant employees and contractors is different due to the silo-focused mentality of the ammonia management and employee teams (see Table 4.5). Contractors are focused on driving economic performance and are reluctant to disclose previous safety incidents. Potential limiting factors observed at the ammonia plant from Table 4.5 are arrogance, image, long hours, blame, and discrimination, which are directly linked to the human factors assessment. Employees and contractors require more resources to issue permits and complain about the quality of training received and the number of tasks executed. Employees do not accept accountability for their work and the safety of contractors as a result of the blame culture.

Values Plot

	Personal Values	Current Culture Values	Desired Culture Values
	IRS (P)= 6-4-0 | IRS (L)= 0-0-0	IROS (P)= 2-1-8-0 | IROS (L)= 0-0-0-0	IROS (P)= 3-4-6-0 | IROS (L)= 0-0-0-0

Matches
PV - CC 0
CC - DC 6
PV - DC 5

Health Index (PL)
PV: 10-0
CC: 11-0
DC: 13-0

#	Personal Values			Current Culture Values			Desired Culture Values		
1	honesty	55	5(I)	*employee safety*	59	1(O)	*teamwork*	39	4(R)
2	commitment	46	6(I)	cost reduction	46	1(O)	*continuous improvement*	37	4(O)
3	family	44	2(R)	profit	44	1(O)	employee recognition	36	2(R)
4	safety	42	1(I)	*productivity*	42	3(O)	*employee safety*	34	1(O)
5	accountability	37	4(R)	*customer focus*	39	2(O)	honesty	32	5(I)
6	respect	37	2(R)	*continuous improvement*	36	4(O)	integrity	29	5(I)
7	balance (home/work)	32	4(I)	achievement	31	3(I)	commitment	28	5(I)
8	dedication	31	5(I)	*quality*	27	3(O)	customer focus	28	2(O)
9	friendship	31	2(R)	competitive	25	3(O)	trust	27	5(R)
10	integrity	30	5(I)	excellence	22	3(I)	accountability	26	4(R)
11				*teamwork*	22	4(R)	balance (home/work)	26	4(O)
12							*productivity*	26	3(O)
13							*quality*	26	3(O)

Black Underline = PV & CC   Orange = CC & DC   P = Positive   L = Potentially Limiting (white circle)   I = Individual   O = Organizational
Orange = PV, CC & DC   Blue = PV & DC   R = Relationship   S = Societal

Values Plot   Copyright 2011 Barrett Values Centre   September 2011

Values Distribution

	Personal Values	Current Culture Values	Desired Culture Values
7	4%	2%	4%
6	4%	4%	8%
5	29%	10%	19%
4	19%	14%	21%
3	17% / 1%	29% / 6%	24% / 1%
2	19% / 0%	7% / 6%	15% / 1%
1	7% / 0%	17% / 5%	7% / 0%
	CTS = 37-19-44 Entropy = 1%	CTS = 16-14-70 Entropy = 17%	CTS = 31-21-48 Entropy = 2%

C = Common Good
T = Transformation
S = Self-Interest

Positive Values
Potentially Limiting Values

Values Distribution   Copyright 2011 Barrett Values Centre   September 2011

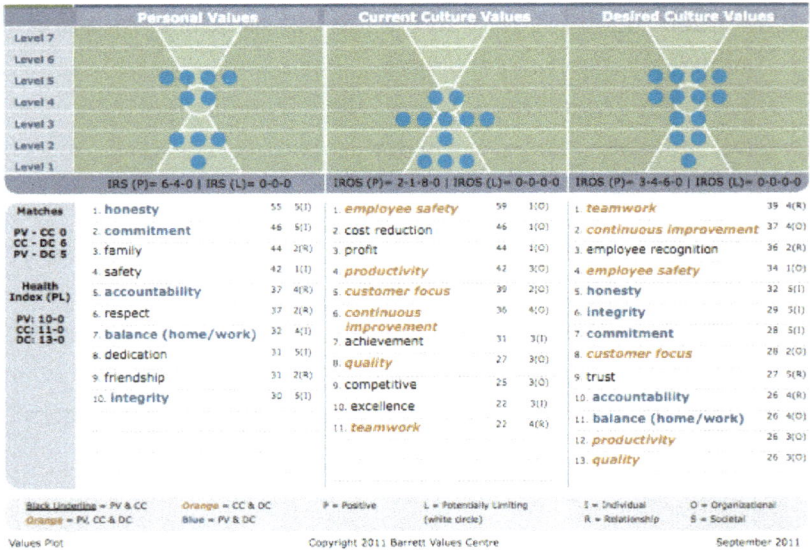

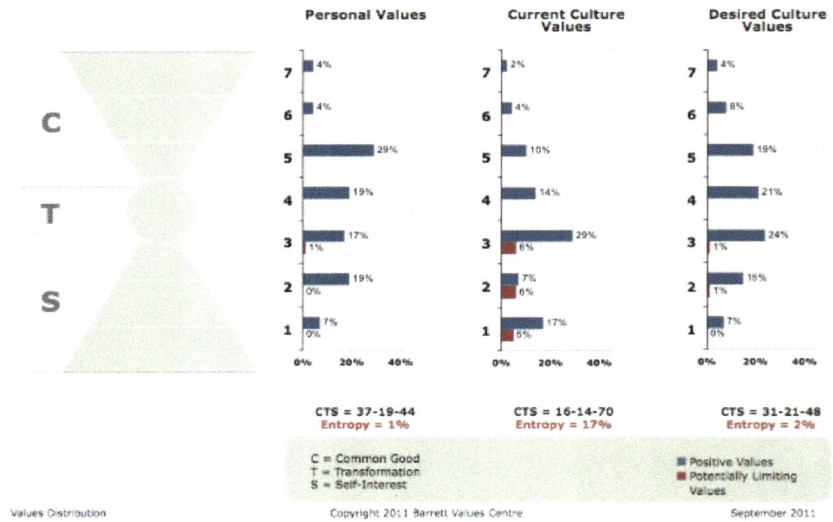

Figure 4.16. Team leadership values results for ammonia plant. Data from Restricted, 2011.

The high first-party audit scores and low incident reporting of hydrocarbon leak FER-SI for the ammonia plant indicate employees are sensitive towards their image and fear shame and blame. The ammonia plant manager also included a zero FER-SI target in all performance contracts, which means employees would feel discriminated against or prejudiced during their performance appraisal if they report process safety incidents. Organizational learning and continuous improvement in the ammonia plant and in other business units are compromised because employees are reluctant to learn from past mistakes due to fear of discrimination and re-crimination when reporting and accepting accountability of incidents.

Table 4.5. Potential limiting values identified in current culture. Data from Restricted, 2011.

Gas-to-Liquid Plant

Level	Potentially Limiting Values (votes)	Percentage Entropy
	image (15) arrogance (10) power (10) bureaucracy (9) hierarchy (7) long hours (5) information hoarding (4) cynicism (2) silo mentality (2)	64 out of 415: 5% of total votes
	blame (16) empire building (13) discrimination (9) internal competition (9) conformity (6) tradition (3) paternalistic (1)	57 out of 167: 5% of total votes
	caution (10) excessive control (9) job insecurity (8) exploitation (4) short-term focus (4)	35 out of 174: 3% of total votes
Total	156 out of 1176	13% of total votes

Steam Utilities

Level	Potentially Limiting Values (votes)	Percentage Entropy
	arrogance (22) image (15) power (14) information hoarding (13) bureaucracy (10) hierarchy (6) long hours (6) silo mentality (5) cynicism (2)	93 out of 353: 8% of total votes
	blame (30) discrimination (22) empire building (10) tradition (9) internal competition (4) conformity (3)	78 out of 213: 7% of total votes
	job insecurity (25) caution (18) short-term focus (14) exploitation (12) excessive control (9)	78 out of 222: 7% of total votes
Total	249 out of 1182	22% of total votes

Effluent and Disposal

Level	Potentially Limiting Values (votes)	Percentage Entropy
	arrogance (17) power (15) information hoarding (13) bureaucracy (11) image (10) silo mentality (10) long hours (9) hierarchy (5) cynicism (4)	94 out of 340: 9% of total votes
	blame (23) discrimination (20) empire building (16) internal competition (8) conformity (7) tradition (6) paternalistic (1)	81 out of 192: 7% of total votes
	job insecurity (18) excessive control (11) short-term focus (9) caution (6) exploitation (6)	50 out of 208: 5% of total votes
Total	225 out of 1105	21% of total votes

Ammonia Plant

Level	Potentially Limiting Values (votes)	Percentage Entropy
	arrogance (14) image (13) long hours (8) power (8) bureaucracy (7) information hoarding (6) hierarchy (4) cynicism (3) silo mentality (2)	65 out of 360: 6% of total votes
	blame (21) discrimination (18) tradition (9) empire building (6) conformity (4) paternalistic (1)	59 out of 130: 6% of total votes
	excessive control (14) caution (12) job insecurity (11) exploitation (9) short-term focus (8)	54 out of 228: 5% of total votes
Total	178 out of 1022	17% of total votes

Summary of Analysis

Organizational process safety maturity related to implementation and sustainability can be increased if there is greater integration of existing management systems, including information management systems, safety communication, revision of SOPs, employee training, MOC execution, reliability engineering, operational excellence, and risk and maintenance management. Dedicated resources are required for compiling SOPs and employee on-the-job training. The Swiss cheese model methodology of providing additional layers of protection, should not be compromised in favor of overemphasizing the current permit to work system. These findings are demonstrated in the human factors perception survey, which were noticed in the downward trends and supported by the human factors assessment interviews. The maintenance integrity audit findings indicate an effective long-term maintenance integrity strategy is required and needs to be aligned to OSHA PSM requirements. A more prescriptive process definition should be made on identifying process safety critical equipment to prevent employee confusion and the duplication of maintenance management systems.

Organizational transformation and effective leadership are required to encourage effective FER or loss of containment reporting when employees feel management bias and their performance is undermined during the incident reporting process. Accordingly, change management is needed to influence and encourage hydrocarbon leak reporting, increase organizational learning, and continuous improvement. The reporting process should not be used as a yardstick for consequence management.

The four plants share common potential limiting factors (see Table 4.5) related to blame, management by fear, image, discrimination, and arrogance, which are the causes of organizational entropy and employee disengagement. The average entropy level across the four plants is more than 18 percent, and the largest potential limiting factors are observed at Level 3 and contribute to an average of seven percent. These behaviors also include the unwillingness to accept accountability for human factors-related findings and recommendations, which was noticed during the feedback session with the respective management teams. An entropy level of more than ten percent suggests organizational restructuring and rapid transformation need to occur to foster high organizational team performance and increase process safety maturity.

There are strong self-esteem (Level 3) behaviors displayed by employees in each department, which is driven by upholding one's departmental and personal image and the fear of loss of reputation and discrimination. However, employees do seek higher maturity leadership behaviors shown in the desired culture focused on transformation (Level 4) and internal cohesion (Level 5). Also, personal or employee safety are noticed across the personal (PV), current (CC), and desired (DC) values dimensions. The absence of "positive" societal leadership behaviors, together with strong personal and organizational self-esteem, suggests that employees are conscious of their own and their immediate colleagues' safety, thus increasing silo-focused behaviors (see Table 4.5). The silo mentality indicates safety and well-being for employees from other plants, and contractors are given less priority.

The lowest scoring human factors survey dimension of staffing and workload is related to the long hours worked by employees in the four plants shown in Table 4.5. Information hoarding is associated with understanding critical processes for specialty gas operations, reluctance to share organizational learning from process safety incidents, and inadequate safety critical communication.

Potential limiting leadership behaviors and overcoming them in the desired culture are identified together with the degree of process safety sustainability deficiencies, and progress made, including assessing incident severity, root causes, and reporting effectiveness. To achieve the accelerated mobilization of process safety culture maturity, all stakeholders must experience safety leadership and witness leadership demonstrating the organizational values of trust and accountability

In addition to investigating underlying leadership behaviors, assessing the outcomes from the human factors assessment survey and interviews with process safety sustainability progress, incident reporting effectiveness, and the root cause analysis allows for the determination of the process safety culture maturity for the four plants and is described in Chapter 5.

CHAPTER 5 PROCESS SAFETY MATURITY ASSESSMENT

Alignment of Maturity Models

The human factors assessment, Loss of Containment FER severity rate, process safety audit findings, and scores are used to assess the safety culture maturity using three models: Eames and Brightling, HSE, and the Bradley DuPont Curve. The models are related to each other and are described in Table 5.1. Each plant is assessed for different safety maturity dimensions using the Eames and Brightling (2012) safety maturity model and then aligned with the remaining two models.

Table 5.1. Alignment of safety culture models

Eames and Brightling, 2012	HSE, 2000	DuPont, 2009
Cognizant	Level 1 (Emerging safety maturity)	Safety managed by Reactive/ Natural Instinct
Informed Reporting	Level 2 (Emerging safety maturity, Managing, commitment, rules, procedures, FER performance implies punishment or reward)	Dependent or Supervisory safety culture
Just Culture	Level 3 (proactive safety involvement by teams with management to improve safety performance)	Self and peer group safety monitoring

Operational Discipline	Level 4 (Staff committed towards safety performance. Get it done first time right. Ineffective management decisions cause incidents. Staff and management are accountable for safety performance and incidents)	Self and team leadership safety. Others keeper or help others conform. Organizational pride
Organizational Learning	Level 5 (No complacency when FERs are zero. Develop consistency and fight failure. Learn from own and other employees' failures without blame, shame, or prejudice. Continuously improve and raise safety standards for others to follow)	Organizational learning and continuous improvement. Employee teams are more accountable for their own safety and raise safety performance standards

Eames and Brightling Safety Culture Assessment

The model proposed by Eames and Brightling (2012) consists of five successive stages of maturity development, including cognizant, informed reporting, just culture, disciplined, and learning dimensions, which are assessed for each department.

Cognizant:

The human factors assessment feedback for the gas-to-liquid plant management shows management, plant operators and maintenance teams are not aligned with their departmental safety strategy and that none of these teams function as an integrated unit. Management commitment to safety is undermined due to reluctance in implementing safety recommendations or by failure to accept human factors findings and acknowledge the need to investigate safety behaviors at the management and floor level further. Plant operators and maintenance teams are enthusiastic about "living" the value of safety and upholding their team and personal image.

Similar to the study conducted at the Synthetic Unit by Payne et al., (2010), employees in all assessed plants do not effectively use SOPs and instead rely on their memory or "gut instinct" (for example, the steam utilities control room employees) to make knowledge-based decisions during plant upsets. Incorrect safety behaviors are demonstrated by not using inspection checklists and updating SOPs after every incident or every two years. Safety collaboration is noticed for gas-to-liquid control room and maintenance teams as employees are motivated when discussing safety problems and proposed solutions.

Informed Reporting:

An effective reporting culture is noticeable at the gas-to-liquid plant since adequate FER incident reporting, root cause identification, and safety management correction were implemented from 2009 till 2012. The steam utilities, effluent and disposal, and ammonia plant management teams instill fear in their employees and encourage a punitive work culture when incidents occur or when the leak FER severity rate increases, despite acknowledging deficiencies in their safety management systems and safety behaviors. Consequently, inadequate reporting and correction of safety management systems are observed in these plants. The first party audits at the ammonia plant demonstrate employees do not feel confident about addressing management system deficiencies due to fear or reputational damage and prefer maintaining the status quo, hence the high scores noticed for all process safety standards, which indicates all system deficiencies have been addressed and no further improvement is necessary. The remaining plants acknowledged second party audit findings but still lack commitment towards improving their safety management systems.

Just Culture:

Process safety implementation and sustainability are conducted in teams for each plant. There can be divisiveness between the management teams and employees regarding safety management. Trust and accountability are replaced with fear and blame in all the plants—especially between the management and workforce, as seen in the Barret survey results.

In addition, the human factors assessment shows employees complain about incident learning from other plants that are not shared due to fear of losing organizational reputation. The assessment also demonstrates plant managers are unable to increase their process safety management system effectiveness after an incident occurs.

Disciplined:

Operational discipline is ineffective for all plants since employees do not rely on SOPs. Instead, they depend on their experience and competencies to make knowledge-based decisions as SOPs exclude troubleshooting guides. Employees are satisfied with the shift hand-over communication, as discussed in the human factors assessments. This acceptance positively contributes to operational discipline since every shift team is aware of the risks in the plant. The consistency of how plants are started, operated, and shut down, however, is ineffective if operational discipline is not managed. SOPs and checklists are only updated after an incident occurs and not during task observations. Employees master operational discipline through on-the-job training, and there is a disconnect with formal theoretical and practical training assessments.

The permit to work process, which is the last layer of protection in the Swiss cheese model, is emphasized and used as a trade-off against multiple layers of protection associated with effective SOP compilation, training, and compliance. Employees comply fully with the permit to work process. In part, this is due to a mindset of fear and blame since permits are audited regularly for compliance of more than 95 percent, and deviances are remedied using the disciplinary management process.

Learning:

Organizational learning is undermined in all plants due to the reluctance to freely share incident lessons among all the plants and how to use that knowledge to reduce safety incidents. Continuous improvement through increasing safety management effectiveness is made mostly by risk assessments and second and third-party audits. Also, plant management needs to communicate safety incidents in a coherent method by making incident lessons and root causes relevant to each plant. Plant operator employees with silo mentalities struggle to understand the relevance of incidents from other plants and how the associated lessons can be implemented in their own plants. None of the employees assessed during the human factors interviews indicated caring, safety, or leadership behaviors for employees from other plants, site contractors, or public stakeholders. These findings are aligned with the results from the Barret survey indicating there are no societal leadership behaviors present in any of the plants.

The analysis demonstrates a cognizant and informed reporting culture is visible in selected plants. The safety maturity and culture are undermined due to organizational blame, fear, and silo mentality, which encourages inadequate reporting of leaks and discourages the sharing of lessons learned from incidents in other plants. Operational discipline is present only for the permit to work process; however, employees need to rely on their experience and knowledge for operating the plant, thus causing operational inconsistencies.

HSE Process Safety Culture Assessment

Figure 5.1 shows the process safety culture model by HSE (2000). The gas-to-liquid plant is in transition from Level 3 to Level 4, indicating employee teams manage their own safety behaviors and are proactive in safety management when developing new safety benchmarks. Employees perceive gas-to-liquid plant management teams as enablers to safety, and more integration is required between employees and the management team regarding safety strategy development and execution.

The steam utilities, effluent and disposal, and ammonia plants are in transition from Level 2 to Level 3 safety culture maturity. Management teams are committed towards safety; however, compliance with safety is driven by employee fear. There is a lack of trust between employees and their management teams, and no one accepts accountability for safety performance and sustainable process safety management. Operational, management, and safety teams work in silos. Management perceives safety teams are accountable for effective safety management systems and safety performance, while operational teams are accountable for process safety lagging metrics, such as hydrocarbon leak trends. A punitive blame culture exists in these plants since hydrocarbon leak trends are used to undermine employee performance instead of encouraging continuous improvement and organizational learning.

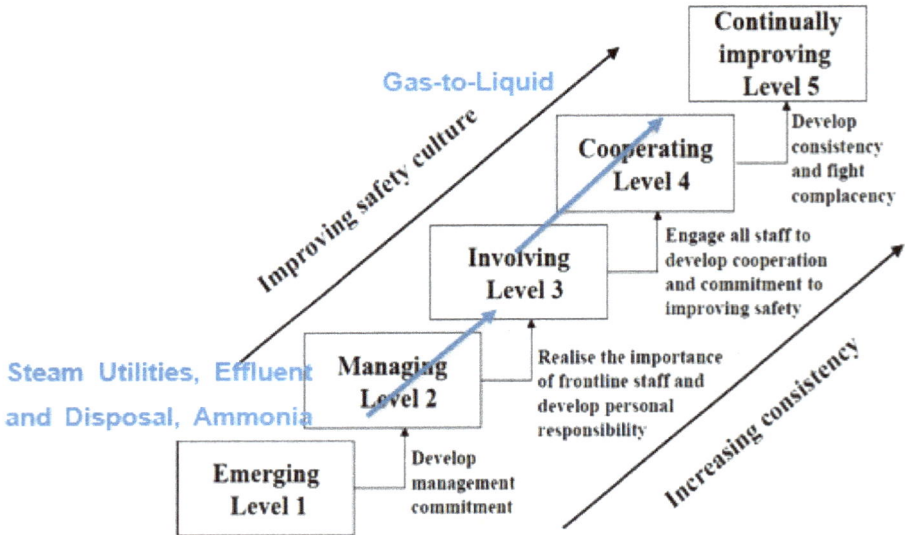

Figure 5.1. Safety cultural maturity levels. Data from HSE, 2000.

A key challenge that undermines the process safety culture maturity from the HSE (2000) model observed in all plants is the lack of clear and consistent safety communication. Examples include:

- Inadequate equipment labeling and the 'touch and tag' of critical equipment concept before work execution.
- Problems with the remote and control room communication protocol (including the use of other languages in addition to English) when critical operations are performed.
- The unavailability and reliability of communication devices. CCTV cameras are not in place to inspect live safety critical equipment and operations. The two-way radios are inadequate,

and the control room panel layout and screen contents cannot be easily read. There are navigational errors when operating SCADA or DCS due to finger slip errors.

- Inconsistent equipment labeling or the touching and tagging of equipment resulted in numerous process safety incidents as contractors and plant employees from different plants misinterpret communication. There is no single procedure that governs and encourages a common understanding of how equipment is labeled or 'touch and tagged' for work purposes throughout all sites. The current labeling format resulted in repeat process safety incidents at the ammonia and effluent and disposal plants.

Safety culture maturity development can be accelerated for all plants with effective leadership development, increased accountability, continuous improvement, organizational trust, and a high degree of clear, consistent safety communication and interpretation. Best practices derived from gas-to-liquid and safety incidents should be freely shared with other plants without shame, fear, or loss of reputation to stimulate safety learning and improve safety management effectiveness.

Plant management needs to realize process safety incident root causes occur due to a wide range of factors originating from management decisions and employees. The plateau in the leak incident severity rate described in Chapter 4 should discourage complacency by employees and management teams. Also, they should nurture behaviors regarding a "chronic sense of unease" since process safety incidents have a high consequence and low probability of occurrence. A Level 5 process safety culture encourages "constant paranoia" that the next process safety incident is imminent, and a high-performance organization is geared towards continuous improvement.

DuPont Safety Culture Maturity Assessment

Figure 5.2 shows the DuPont safety culture maturity with associated process safety and human factors dimensions. The organization's process safety experts selected appropriate dimensions and safety culture maturity scales to assess process safety maturity.

Second and third party process safety audit scores are included, which demonstrate employee commitment and the effectiveness of process safety management systems. The adequacy of hydrocarbon leak severity rate disclosure, the frequency of incident category, and room for continuous improvements, based on new learning implemented in all plants from process safety incidents, are assessed together with departmental entropy of current culture values, derived from the Barret survey and total scoring of the human factors perception survey per plant.

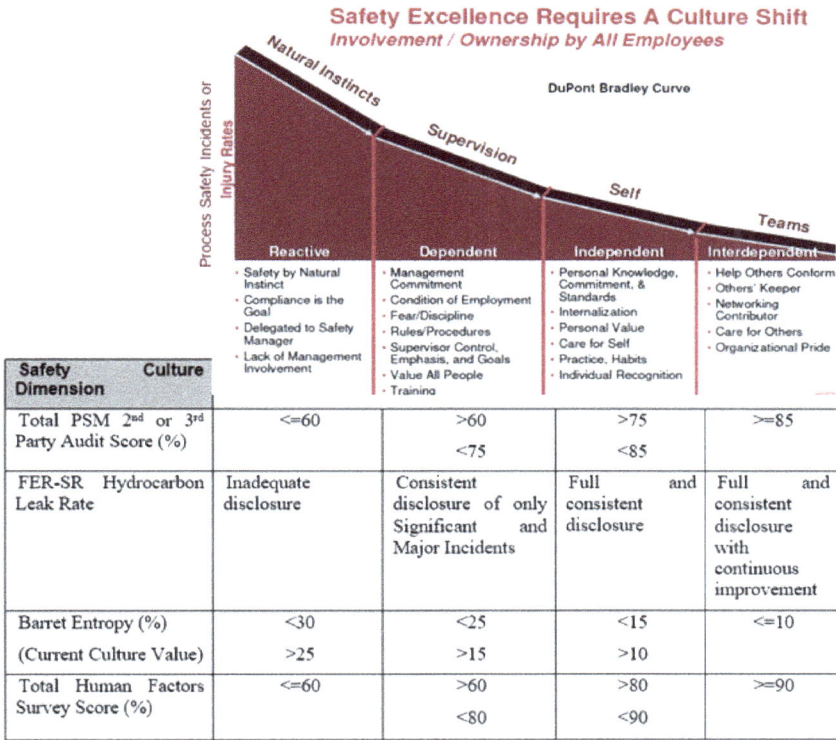

Safety Excellence Requires A Culture Shift
Involvement / Ownership by All Employees

DuPont Bradley Curve

Safety Culture Dimension	Reactive	Dependent	Independent	Interdependent
	· Safety by Natural Instinct · Compliance is the Goal · Delegated to Safety Manager · Lack of Management Involvement	· Management Commitment · Condition of Employment · Fear/Discipline · Rules/Procedures · Supervisor Control, Emphasis, and Goals · Value All People · Training	· Personal Knowledge, Commitment, & Standards · Internalization · Personal Value · Care for Self · Practice, Habits · Individual Recognition	· Help Others Conform · Others' Keeper · Networking Contributor · Care for Others · Organizational Pride
Total PSM 2nd or 3rd Party Audit Score (%)	<=60	>60 <75	>75 <85	>=85
FER-SR Hydrocarbon Leak Rate	Inadequate disclosure	Consistent disclosure of only Significant and Major Incidents	Full and consistent disclosure	Full and consistent disclosure with continuous improvement
Barret Entropy (%)	<30	<25	<15	<=10
(Current Culture Value)	>25	>15	>10	
Total Human Factors Survey Score (%)	<=60	>60 <80	>80 <90	>=90

Figure 5.2. DuPont Bradley Curve Safety Culture Model. Data from DuPont, 2009.

Table 5.2 shows the safety culture assessment created for the four plants. The steam utilities plant is in transition from a reactive to a dependent cultural state despite effective process safety implementation. This condition suggests long-term process safety sustainability may be undermined due to ineffective leadership behaviors. Organizational restructuring occurred at the steam utilities plant after the second party process safety audits were conducted during 2009-2010, indicating team leadership behaviors and human factors issues, including employee training, staffing and workload, incident reporting, and investigation, deteriorated during 2010 until the present, as seen in Figure 3.1 and the Barret current culture entropy score.

Table 5.2. Process safety culture assessment using DuPont Bradley Curve

Safety Culture Dimensions	Reactive	Dependent	Independent	Interdependent
Total PSM 2nd or 3rd Party Audit Score (%)	None	Gas-to-Liquid (69) Effluent and Disposal (68)	Steam Utilities (80)	None
Hydrocarbon Leak FER-SR	Steam Utilities	Effluent and Disposal	Gas-to-Liquid and Ammonia	None
Barret Entropy (%) (Current Culture Value)	None	Effluent and Disposal (21) Ammonia (17) Steam Utilities (22)	Gas-to-Liquid (13)	
Total Human Factors Survey Score (%)	Steam Utilities (55)	Ammonia (66) Effluent and Disposal (63)	Gas-to-Liquid (86)	None

Note: Ammonia Plant first-party PSM audit result is excluded from this assessment due to employee bias.

The effluent and disposal plant is in a dependent safety maturity state, indicating management is committed towards safety. At the same time, employees are subjected to continual supervision, discipline or rules, creating fear, a lack of trust, and a blame culture. These behaviors already surfaced in the Barret survey analysis. They showed a lack of accountability, long hours, arrogance, blame, and human factors dimensions related to compliance with operating procedures, manpower planning, ineffective visible felt safety leadership. In part, this occurs since employees perceive management as only seeking work productivity without addressing safety concerns during plant inspections.

The ammonia plant is in transition from a dependent to an independent safety culture following organizational restructuring. Consistent leak severity rate disclosure was announced in January 2012. Yet there is still noticeable excessive control, blame, discrimination, and fear of loss of reputation since employee performance is undermined whenever hydrocarbon leaks are disclosed or if process safety audit scores are less than benchmarked levels. The human factors survey and interviews conducted by ERM (2007) indicate excessive supervision and the quality of procedures and employee competency assurance deteriorated with time, as confirmed in Figure 3.1. In 2007, ammonia plant contractors' employees feared the loss of their bonuses if they reported safety incidents. Also, employees overrated their process safety management first-party audits (see Figure 4.1) to uphold the plant's image and avoid discrimination. The independent or "self" safety leadership maturity emerging at the ammonia plant is due to the introduction of behavioral safety monitoring of fellow employees and the organization's safety management commitment and philosophy.

The gas-to-liquid plant is in an independent or self-leadership maturity state based on leadership, human factors survey, and FER disclosure rates. There are inadequate resources allocated to implementing and sustaining an effective process safety management system, especially maintenance integrity. Employees are, however, committed to improving safety performance, as demonstrated by the human factors assessment feedback. The plant management team and employees do not have close integration and cooperation regarding safety strategy and execution due to the silo mentality, image concerns, arrogance, and blame, as disclosed from the Barret survey.

An interdependent culture is not fully developed in all the plants since employees, management teams, and contractors encourage silo-focused behaviors, and safety performance is measured per plant instead of reviewing consolidated safety performance indicators. The Barret survey and human factors assessment also indicate the absence of caring, societal safety leadership behaviors for employees from other plants or the community. Employees driven by fear and blame develop a "self-protectionist" mindset regarding personal and process safety behaviors, undermining safety team development across different plants.

Summary of Process Safety Culture Assessment

A cognizant culture is observed in all plants regarding operational discipline and knowledge-based decision-making. Employees do not know the significance of SOPs and mostly rely on their experience and judgment when making operational knowledge-based decisions, especially during plant troubleshooting. Also, the permit to work system removed the spotlight from operational discipline due to plant management teams emphasizing compliance to the permit to work principle (as disclosed from PSM audit scores). This behavior occurred instead of transforming employee culture to one that verifies effectiveness, adherence to SOPs, and the creation of multiple protection layers of defense to prevent safety incidents. Operational discipline only is considered after an incident occurs and triggers the revision of checklists and SOPs instead of conducting critical task observations, thus encouraging incorrect process safety behaviors.

An informed and reporting culture exists in the gas-to-liquid and ammonia plants. There are, however, still elements of blame, fear, and loss of reputation experienced in all plants that weaken an effective reporting culture striving towards continuous improvement and organizational learning.

The HSE safety maturity model indicates that the steam utilities, effluent and disposal, and ammonia plant management teams are committed towards safety. Yet, employees need to be supervised through rules, procedures, and fear, thereby encouraging "dependent" employee safety behaviors. Employees believe management is responsible for their safety through adherence to procedures, yet there is a lack of demonstrated effective team leadership safety behaviors in these plants. Gas-to-liquid plant management and their employees behave as two divided teams, thus compromising safety performance and strategy execution, despite both teams being committed to improving the process safety performance and sustainability. Plant management teams should increase employee engagement by including all employees when addressing accountability and developing and executing new safety strategies. To accelerate safety maturity, "authentic discussions" need to occur regarding the value added by rules, procedures, and operational discipline versus required safety behaviors.

Conclusions related to the outcomes from the human factors and safety culture assessment for each plant and how to fast track process safety maturity through effective man-machine, job roles and organizational interfaces, supported by desired leadership behaviors, are discussed in Chapter 6 along with how human factors and maturity assessments should be managed in the future.

CHAPTER 6 CONCLUSIONS AND RECOMMENDATIONS

CONCLUSIONS

Conclusions are derived from the role of human factors dimensions associated with safety culture and the identification and management of associated risks, together with the status of the current process safety maturity. The identification of human factors dimensions that impact process safety sustainability, driven by potential limiting leadership behaviors and development areas, are discussed.

Role of Human Factors in Safety Culture

The study of human factors and safety culture was developed due to global and local process safety incidents, which caused business interruption and loss of shareholder value due to organizational reputation damage. The human factors study reflects leadership behaviors are the primary drivers of crucial human factors strengths. These include alarm management and priority, behavioral safety and change management and weaknesses or underperformance in areas of operational discipline, staffing overload, employee competence, supervision, culture, and leadership. Process safety incidents occurring in the broader organization largely contributed to human error and ineffective leadership supported by incorrect cultural values. During 2011, a change was made in the organization's values, which included accountability, increasing simplicity without bureaucracy, breaking down silos, and providing clear and consistent communication. In part, these new leadership values contributed to a reduction in process safety incidents reported in the first and fourth quarters of the financial year 2012. Employees and contractors still need to internalize and demonstrate these values to achieve sustainable reductions in process safety incidents.

Risks Derived from Human Factors

Some of the high level 1 risks identified in the human factors assessment and survey that contributed to incidents include non-existent or non-compliance with SOPs, incorrect equipment and pipe labeling, and ineffective safety culture and leadership behaviors. Some of the medium levels 2 and 3 risks are inadequate maintenance, ineffective safety communication, alarm management, safety systems, and insufficient manpower during shifts. The risk management framework is used to prioritize the allocation of resources and to sensitize management and employees regarding the seriousness of addressing deficiencies undermining process safety sustainability. The results from the human factors assessment and survey are aligned with the contributing factors when addressing site-wide process safety incidents and reduction in hydrocarbon leak rates.

Process Safety Maturity Assessment

Three process safety maturity assessment models derived from Eames and Brightling (2012), leadership effectiveness using HSE (2000) and the DuPont Bradley (2009) curve were applied to each plant-based upon operational discipline, incident reporting, and organizational learning using process safety balanced scorecard metrics. Eames and Brightling (2012) discussed the five successive stages of maturity development: cognizant, informed reporting, just culture, disciplined, and learning dimensions, which all were assessed for each plant.

A cognizant culture is seen at the steam utilities, ammonia and effluent, and disposal plants. This environment exists because employees lack operational discipline and perceive process safety sustainability as a continuous struggle without success. There is too much reliance on employee experience and knowledge instead of using operating procedures or troubleshooting guides to aid in complex start-up decision-making at the plant. Organizational learning maturity and continuous improvement are undermined because employees fear management and are reluctant to escalate process safety incidents or share lessons from past mistakes, thus repeating incidents related to working on operational equipment. The organizational maturity is not geared to continuous risk reduction as a function of process safety improvement and increasing legislative stringency.

A just and informed reporting culture is implemented with mixed success at the gas-to-liquid plant. Employees do not fear reporting process safety incidents and FER-SI or FER-SR scores monthly. The remaining plants, however, are unwilling to effectively report incidents due to a manifestation of a punitive work performance culture leading to employees feeling discriminated against whenever process safety incidents are escalated. A culture of fear and blame exists in the organization, which prevents it from institutionalizing organizational learning and continuous improvement. Employees in all plants struggle with the interpretation of process safety incident learning outcomes because they are either unable to find the relevance of the lessons shared, need coaching, or cannot discuss learning in their plants.

A disciplined culture as a key strength lies in the permit to work process, which is the last layer of protection before any work execution. However, there is no emphasis on effective employee competence assurance or on relying on operating procedures. Operational discipline strength includes the change management process related to equipment or hardware; however, organizational and legal changes are ineffectively managed in all plants.

The HSE (2000) maturity assessment indicates the gas-to-liquid plant is in transition from recognizing the importance of plant employees and demonstrating the organizational values, to engaging all staff to commit to safety improvement and effectiveness. The remaining plants are in transition from adequately demonstrating safety management commitment towards recognizing the importance of employees playing a significant role in effective safety management without fear or blame. More integration between front-line staff and management teams, HSE motivational discussions, and collaboration are needed in all plants. When this happens, it will accelerate safety maturity and the removal of fear, blame, the silo mentality, and the upholding of the organizational and personal image—all of which are visible during process safety incidents and audits.

The DuPont Bradley (2009) Curve and associated process safety metrics are used to assess effective team leadership behaviors measured as a function of process safety audits, hydrocarbon leak rate FER-SR, Barret entropy scores, and human factors survey scores. The steam utilities plant is in transition from a reactive to dependent safety culture. This represents a shift from a situation where employees drive safety performance by using intuition when operating equipment to one where operating procedures are used for risk-based decisions making. There also is a lack of plant management commitment to rules and supervision-based safety with an emphasis on employee coaching. The effluent and disposal plant is in a dependent state, indicating the management's strong safety commitment. Employees, however, are subjected to rules and procedures driven by a lack of trust, blame, and fear, and employees struggle with making decisions concerning production demand requirements versus addressing process safety risks.

The ammonia plant is transitioning from a dependent to an independent safety culture due to the recent merger within the organization. There are elements of blame, fear, lack of trust, and excessive control whenever process safety audits need to be executed or when process safety incidents occur, and an increase in effective reporting needs to be emphasized. The merger allowed the ammonia plant to introduce behavioral safety monitoring, which encouraged strong safety habits and caring for fellow colleagues. A valuable practice recognized at this plant is that the DuPont Bradley Curve is discussed and monitored every month to identify problem areas and room for improvement. An independent safety culture exists at the gas-to-liquid plant since employees are responsible for their own safety, and they demonstrate the safety value through strong behavioral safety management, transparent incident reporting behaviors, and their willingness to learn and continuously improve. Gas-to-liquid management and front-line employees are, however, divided and do not execute a common safety strategy, which is preventing this plant from maturing to a unified safety team.

The results of the three assessments demonstrated the gas-to-liquid plant did the best in process safety maturity, followed by the ammonia, effluent and disposal, and steam utilities plants. Common potential limiting leadership behaviors, as a function of Barret entropy scales and low human factors survey scores, are predictors that undermine mobility to the next maturity level.

Effective Process Safety Sustainability with Human Factors

The human factors assessment indicates there are strengths in the organization related to risk management, alarm management, behavioral safety, and the management of hardware changes. These strengths are effectively sustained to enable process safety performance.

Identified areas of weaknesses that resulted in process safety incidents include the lack of operational discipline and employee competence, ineffective organizational learning, inconsistent safety communication, non-compliance to maintenance frequencies, incorrect control room layout, inadequate manpower levels, and lack of appropriate supervision. Other areas of weakness are the culture and leadership applied to process safety, as demonstrated in the human factors assessment and the site-wide process safety incident analysis. Also, the lack of organizational learning weakened process safety sustainability due to employees' reluctance to share and implement lessons from incidents, thus increasing the likelihood of repeat incidents.

Management's response to the human factors assessment suggests the gas-to-liquid management and plant employee teams behave in silos, and there is a lack of management commitment towards addressing any human factors recommendations. On the other hand, effluent and disposal and steam utilities plants welcome and agree with the outcomes of the findings and recommendations of this assessment. Still, plant managers prefer to interrogate their management teams without accepting accountability. Management teams use excessive control due to the over-reliance on compliance with procedures instead of investigating the lack of operational discipline.

Plant management responses to the human factors assessment and the lack of leadership synergy between management teams and front-line employees indicate an interdependent safety culture focusing on continuous improvement and organizational learning is in jeopardy without leadership development. The success of addressing human factors weaknesses and risks to enable effective process safety sustainability is determined directly by leadership maturity and development on an individual and team basis.

Primary Drivers for Leadership Development

The assessment indicates a potential limiting factor affecting safety maturity acceleration is the silo mentality or inadequate societal leadership identified in the Barret survey. This condition is represented by employee teams that do not entirely consider the safety wellbeing of contractors and by management teams that do not adequately communicate or engage with front-line employees regarding process safety risks. Polarization in safety management performance is noticeable in contractors, plant employees, and plant management. An adversarial mentality, when developing and executing safety strategies, is evident with contractors, management teams, and front-line employees.

A lack of trust, as well as the existence of blame, image concerns, and fear, are present in all plants. As a result, employees do not report incidents, make safety suggestions, or compile and contribute towards robust HSE strategy development and execution since they feel threatened by management. Ineffective process safety incident reporting and not sharing lessons learned are indicators of fear and a lack of trust when employees believe management is biased or unfair when conducting performance appraisals that could result in disciplinary action. The Barret survey indicates there is excessive management control in favor of managing safety instead of allowing employees to internalize safety values and develop their own procedures and methods for compliance.

Employees misperceive that every non-compliance deviation in operating procedures, especially in the permit to work process, will force disciplinary action, thereby increasing employee fear. In what may be considered a controversial position, management teams believe that only gross violations of work procedures resulting in catastrophic impacts, as described in the risk management framework, will initiate disciplinary proceedings. In addition, management misconstrues hydrocarbon leak rate reporting will create the loss of organizational reputation, instead of assessing long-term process safety trends and analyzing root causes to address continuous improvement. Leadership development is necessary in the areas of team and employee engagement, safety performance appraisals and disciplinary process coaching, societal leadership, and the internalization of organizational values to allow for the acceleration in process safety maturity. The findings indicate process safety culture assessment and sustainability through human factors dimensions are fortified by leadership effectiveness and constructive employee engagement.

Summary of Conclusions

Enabling sustainable process safety management through safety culture impact assessment using human factors dimensions was demonstrated effectively in this assessment by identifying key strengths, weaknesses, and risks that directly influence safety maturity. Some of the critical success factors that can accelerate process safety maturity are organizational learning and continuous improvement supported by interdependent team leadership behaviors. The maturity scales measured in the plants as a function of management commitment, Barret entropy (employee disengagement) level, and human factors survey scoring revealed that the gas-to-liquid plant did the best, followed by the ammonia, effluent and disposal and steam utilities plants. Each plant has a standard set of potential limiting factors related to blame, fear, silo mentality, excessive management control, image concerns, and discrimination (toward other teams and contractors). These behaviors weaken leadership development and safety maturity. Also, employees struggle to practice the organizational value of accountability and do not provide consistently clear communication, which hinders the execution of continuous improvement safety strategies and impedes the transition of the process safety maturity to the next level.

Recommendations

There are ways to accelerate the development of a strong process safety culture. They include recommendations derived from weaknesses identified in human factors dimensions applicable to human-machine interfaces. Also, they are based on mitigating the interaction with employee job roles and organizational climate.

Addressing Human Factors Deficiencies

Conduct an ergonomics study and compile procedures to address the minimum distances required between equipment, equipment labeling, and employees working in confined areas or at elevated heights who need to carry tools and machinery. This can be accomplished without conducting a risk assessment. Pumps are located close to each other at the steam utilities plant, and scaffolds are located directly next to live pump switches at the effluent and disposal plant, which could result in the inadvertent operation or isolation of equipment. Affected employees can inadvertently damage live equipment or even perform maintenance work on the wrong equipment due to equipment similarity or insufficient spacing between equipment.

Make provisions for dedicated resources in each plant to revise SOPs according to process safety standard requirements instead of using the permit to work system as the only layer of protection. Add equipment labels or identification to operating procedures and include critical task observations as an "add on" to the existing occupational behavior safety techniques monitoring program. Create a database to capture slips, lapses, violations, and mistakes in process safety critical task operations. Use the information from this database to increase human reliability and reduce human error in revised operating procedures.

Maintenance resources are based on the work priority level. In certain instances, however, lower priority maintenance issues are not addressed as early as possible, in spite of resource availability. Develop a productive time and maintenance priority schedule between the gas-to-liquid plant and the maintenance provider to ensure all maintenance work in specific areas is completed within planned time limits without affecting pre-startup safety activities or safety critical equipment. Include in production shift manpower levels an extra day for training as the current shift system varies for all four plants. Selected shifts and employees who were transferred or on leave do not receive a monthly training day to refresh operational discipline. All plants require safety employees to be visible on the plant and to identify and manage risks. Provide training to affected employees on how to identify and manage process safety risks.

A checklist or "tick the box" mentality is exhibited in all plants when compiling monthly alarm trip lists since alarm priority trip trends, spurious trips, and alarms are not investigated and rectified, especially for medium to lower priority alarms. Operators usually deactivate high-priority alarms, without immediately informing the production manager. Process safety near-miss incidents frequently occur, and operators fear reporting such alarm trips. To rectify the situation, make sure that high-priority alarm deactivation communication is automatically sent to the production manager, and an incident investigation is initiated as soon as possible to identify and remedy the incident's root cause. Moreover, trends for medium to low priority alarms should be investigated, and root causes rectified.

Inconsistent safety communication resulted in employees working on live or energized equipment, causing repeat process safety incidents throughout the organization. Contractors' employees are confused with multiple equipment 'touch and tag' philosophies introduced in each plant, including pipe color coding. For instance, the chemicals sites associate red pipelines for flammable hydrogen gas, while a red color is used to identify firewater pipelines in nearby sites. Other communication challenges include equipment that was identified or tagged as a new plant extension where no work is to be conducted, compared to using tags when work needs to be conducted in other areas. Other plants require visual confirmation to identify isolated equipment before conducting work instead of only using the 'touch and tag' principle. A standard site procedure and associated training are recommended to introduce uniform pipe color coding and tagging of equipment to avoid confusion prior to work execution.

Also, two-way radio communication per work discipline and per channel are required to avoid language confusion, along with a communication protocol to confirm the equipment's operational status and the location of field operators. There is a need for a CCTV risk and technology assessment for monitoring high-production process safety critical equipment. This is necessary because operators are unable to continuously monitor the affected equipment for fires, explosions, and releases due to limitations of the existing infrastructure. Reaction time during a catastrophic process safety incident is critical to minimize loss of life and damage to equipment.

Control panel layout effectiveness can be improved for all plants concerning additional infrastructure and panel layout content. All plants have a high-quality logical structure of the production process, which reflects reality. Additional protection layers are required using process parameter confirmation methodologies to prevent inadvertent "mouse clicks" navigational errors and incorrect operating parameter entries. Control room operators acknowledged that process safety near-miss incidents occurred due to navigational errors. Safe operating ranges that do not rely only on alarm conditions and trip systems are required.

Human factors templates for use in any energy generation organization are available in the Appendices.

Mobilization of Process Safety Maturity

The organization needs to create mechanisms to enable process safety maturity to incorporate continuous improvement and organizational learning supported by interdependent team leadership. One strategy to use is the rotation or transfer of employees to other plants. For instance, high safety performance employees from the gas-to-liquid plant can be transferred to the steam utilities plant, and effluent and disposal plant employees can be sent to the gas-to-liquid plant to accelerate the upward safety maturity mobilization. Both safety and plant employees can be transferred to ensure safety blind spot risks are removed, and continuous improvement is fostered in all plants.

A safety improvement plan (SIP) was developed to aid in plant-specific process safety maturity with limited success. Plant managers are continually criticizing the SIP because managers lack facilitation skills or believe the SIP is not a core business process function and, therefore, will "do the least to get by." The SIP allows for authentic, facilitated plant discussions weekly after reviewing global process safety incidents, and questions are directed to encourage a bias for action implementation towards sustaining process safety. The effectiveness of the SIP rollout is measured by the number of actions identified and addressed. It is suggested that safety leadership psychologists aid each plant in the SIP roll out. This can be accomplished by coaching plant managers on effective facilitation skills and making sure the executive committee members insist on including the execution outcomes of the SIP in the plant managers' performance contracts key performance indicator.

Plant inspections done by line managers shifted toward monitoring occupational safety and plant reliability instead of demonstrating a balanced approach toward asset integrity management, process safety, and man-machine interfaces. Experience proves it is easy to identify occupational safety management deficiencies derived from inspections. At the same time, it is challenging to identify process safety risks due to not adequately reading and understanding operational procedures and being unable to identify the correct process safety behaviors, such as operating the wrong equipment, operating too fast or too slow, and correctly interpreting gauge instruments. Too often, executive committee members and plant managers are unable to ask their operations teams the right questions during field inspections to aid in effective process safety management sustainability. Therefore, it is recommended that all managers undergo procedural training to increase or refresh their operational knowledge and human critical task analysis training to understand various failure modes in human reliability.

Process safety incident discussions, the role of leadership in process safety, and organizational learning can be rolled out at two levels. Process safety stakeholder alignment sessions are suggested for senior management to ensure consensus is reached regarding process safety management system effectiveness, the development of high-level strategies addressing procedures used for equipment identification and execution of process safety critical activities. The second session targets front-line employees and consists of assessing the effectiveness of implemented actions recommended by audits and incident investigations, presenting checklists for use when doing process safety field visits focused on asset integrity, and demonstrating strong leadership and management commitment towards sustaining process safety.

Process safety performance is assessed using lagging indicators, including reviewing the number of process safety incidents, monitoring overdue equipment that was not maintained, or the number of change management projects that are not fully completed. Lagging indicators are predictors of past failures and do not adequately assess the organization's process safety strategy execution and sustainability. Leading indicators provide an effective predictor of current and future process safety performance and evaluate the health of the safety management system. Examples of leading indicators include the number of activated protective devices, number of lifted relief valves, number of uninspected critical equipment, number of process hazard analysis assessments, and audit recommendations that were discussed with plant managers and fully implemented. The forward-looking leading indicators increase the robustness of the process safety strategy execution and lower the probability of process safety incidents occurring.

The success of mobilizing process safety maturity to an advanced level is dependent on the adequacy of identifying and managing all maturity gaps related to man-machine interfaces, employee job roles, and willingness and commitment of the organization to undergo cultural change. The primary driver for accelerating process safety maturity is encouraging effective interdependent team leadership behaviors while eliminating potential limiting behaviors that currently undermine safety maturity transformation.

Institutionalizing Leadership Behaviors

Safety visible leadership development training for managers is recommended to improve behaviors regarding on-site process safety inspections, commitment from managers towards implementing process safety recommendations, and demonstrating caring behaviors towards contractors and employees. In addition, personal insight workshops, held with multi-disciplined employees at different organizational layers and facilitated by a behavioral psychologist, are recommended to ensure participants can demonstrate the organizational values and determine underlying reasons and removal of potential limiting behaviors.

Convene free dialogue sessions with front-line employees and executive team members in a safe and caring atmosphere, where employees can discuss issues of blame, fear, discrimination, and difficulties in executing their job roles, especially those applicable to operational discipline, incident reporting and implementing audit recommendations. These sessions aim to rebuild trust, foster caring relationships, and create teamwork between managers, contractors, and employees. The new value of accountability and how to demonstrate it also must be discussed during the open dialogue and personal insight workshop sessions.

Recognize management and employee teams for continuous improvement in process safety, demonstration of societal leadership behaviors, increased organizational learning, and effective incident reporting. Reward teams from various plants for taking ownership when incidents occur and develop strong platforms for their voices to be heard to ensure other plants do not repeat process safety incidents. These platforms also can prove how well employees have developed their skills and implemented new techniques to manage process safety effectively.

Process safety sustainability through effective leadership is the key driver for improving human factors dimensions and accelerating organizational safety culture maturity. The leadership behavior interventions serve to reaffirm management's commitment to process safety and create an organizational shift where front-line employees, production supervisors, and operations managers are responsible for sustaining process safety, and the Safety, Health, Environment, Risk and Quality department is recognized as enablers of process safety management effectiveness.

Recommendations for Future Work In Human Factors

Execute the human factors assessment every three years or whenever process safety standards are revised, and there is sufficient progress made in addressing process safety audit findings and effectively managing leading indicators. Regular assessments demonstrate the direction in which the organization is headed in terms of organizational safety maturity and identifying and managing gaps in maturity levels. Include contractors and staff from other sites in personal insight workshops, and open dialogue sessions to remove silo mentality behaviors, encourage identification of new best practices and institutionalization of organizational values.

List of Tables

List of Figures

List of Acronyms and Abbreviations

DMS   –   Document Management System

ERP   –   Emergency Response Planning

FER-SI –   Fires, Explosions, Releases Severity Index

HSE   –   Health Safety Executive (UK Legislation)

MOC –   Management of Change

OSHA –   Occupational Safety Health Association (American Legislation)

PSCE –   Process Safety Critical Equipment

PSI   –   Process Safety Information

SOP   –   Standard Operating Procedure

REFERENCES

API 754. 2016. Process Safety Performance Indicators For the Refining and Petrochemical Industries. 2nd. ed. American Petroleum Institute.

Alp E. 2015. Safety Culture and Process Safety. Istanbul, Turkey: II International Process Safety Symposium.

Baker J.A. 2007. The report of the BP US refineries independent safety review panel. United States: Chemical Safety Board (CSB).

Barret, R. 1998. Liberating the corporate soul-building a visionary organization: Seven levels of leadership consciousness. Boston: Butterworth Heinemann.

Baybutt, P. 1997. Human factors in process safety and risk management: Needs for models, tools and techniques. Columbus: Primatech, Inc.

CCPS. 1994. Implementing Process Safety. United States: Centre for Chemical Process Safety, American Institute of Chemical Engineers.

CCPS and Energy Institute. 2018. Bow Ties in Risk Management, A Guidance Document. United States: Centre for Chemical Process Safety, American Institute of Chemical Engineers.

CCPS. 2009. Assessing process safety fires, explosions and toxic releases using FER-SI. United States: American Institute of Chemical Engineers, Centre for Chemical Process Safety.

Cooper, M.D. 2000. "Towards a model of safety culture." Safety Science 36: 111-136.

CSB. 2006. "Fire at Formosa Plastics Corporation: Evaluating process hazard." Chemical Safety Board Doc No: 2006-01-TX. June, USA, 2.

Cullen, L. and Anderson, M. 2005. "Human factors integration for a new Top Tier COMAH Site optimizing safety and meeting legislative requirements." Journal of Process Safety and Environmental Protection 83, no. B2 (March): 101–108.

Cullen, W.D. 1990. The public enquiry into the Pipe Alpha Disaster: Department of Energy. London: HMSO.

Dalijono, T., Lowe, K. and Loher, H.J.O. 2005. "Development and verification of a new approach for operator action analysis." Process Safety and Environmental Protection 83, no. B4 (July): 331–337.

Daniels, A. 2017. "Attributes of Safety Leading Indicators." https://www.aubreydaniels.com.

DuPont. 2009. Overview of DuPont's safety model and sustainability initiatives. USA: DOE and DuPont Safety Initiative.

Eames, P. and Brightling, J. 2012. "Process safety in the Fertiliser Industry a new focus - Ammonia Technical Manual." Paper presented at the 56th Ammonia Safety Symposium of the American Institute of Chemical Engineers, Montreal, Canada, September11-15,2012.

Ergo Plus. "A Short Guide to Leading and Lagging Indicators of Safety." 2017.*http://ergo-plus.com/leading-lagging-indicators-safety-performance/*.

Erickson, J.A. "Corporate culture: The key to safety performance." EHS Today. Accessed March 2007. http://ehstoday.com/news/ehs_imp_33155/.

ERM. 2008. Process Safety Management. Manchester, UK: ERM-Sheffield University.

ERM. 2007. Human Factors Report for Nitro Ammonia Division, South Africa. Manchester, UK: Human Factors and Safety Psychology Department.

Fennel, D. 1998. Investigation into King Cross underground fire. London: Department of Transport, HMSO.

Fisk, A.D., Ackerman, P.L. and Schneider W. 1987. Automatic and controlled processing theory and its applications to human factors problems. Human Factors Psychology. Hancock P.A. ed. North Holland: Elsevier Science Publishers.

Foord, A.G. and Gulland, W.G. 2006. "Can technology eliminate human error?" Journal of Process Safety and Environmental Protection 84, no. 3 (May): 171–173.

HRA. 2008. Human reliability associates: Human factors course notes. UK: Sheffield University.

HSE. 2010. Process safety element standards. United Kingdom: Health and Safety Executive Publications. Accessed January 2010. www.hse.gov.uk.

HSE. 2009. Human factors inspectors checklist. UK: Health and Safety Executive Publications.

HSE. 2000. Safety culture maturity model. Offshore technology report 2000/049. Edinburgh, UK: Keil Centre for HSE.

HSE. 1999a. Development of a Business Excellence Model of Safety Culture. UK: Health and Safety Executive Publications.

HSE. 1999b. Reducing error and influencing behaviour HSG48. Health and Safety Executive Publication. 2nd. ed. Norwich, UK: HMSO.

HSL. 2002. Safety culture: A review of the literature. HSL Publication. UK: Human Factors Group, Crown.

Hudson P. 2000. Safety Management and Safety Culture: The Long, Hard and Winding Road. The Netherlands: Centre for Safety Research, Leiden University.

Khader-Abu, M.M. 2004. "Impact of human behaviour on process safety management in developing countries. Process Safety and Environmental Protection." American Institute of Chemical Engineers 82, no. 6 (November): 431–437.

Kiddam, K. and Hume, M. 2012. "Analysis of equipment failures as contributors to chemical process accidents. Process Safety and Environmental Protection." American Institute of Chemical Engineers, (February): Vol 3-12.

Kletz, T. 2006. "Human factors and management." Process Safety and Environmental Protection 84, no. 3: 159–163.

Knegtering, B.A. and Pasman, H. 2009. "Safety of the process industries in the 21st century: A changing need of process safety management for a changing industry." Journal of Loss Prevention in the Process Industries , (November): Vol 162–168.

OSHA. "Process Safety Element Standards." Occupational Health and Safety Association and Department of Energy Publication. Accessed: January, 2010. http://www.osha.gov.

OSHA. 1993. Process safety audit protocol. Process Safety Element Standards S1.14. USA: Occupational Health and Safety Association and Department of Energy Publication.

Parker, D., Lawrie, M., and Hudson, P. 2006. "A framework for understanding the development of organizational safety culture." Safety Science 44, no. 6: 551-562.

Payne, C.S., Xu, V.X., Bergman, M.E. and Beus, J.M. 2010. Process safety culture project. Research report., South Africa and Department of Psychology at Mary Kay O'Connor Process Safety Centre, Texas A and M University, June.

Restricted. 2012a. FER trends and causes: Organizational process safety workshop. South Africa: HSE Corporate.

Restricted. 2012b. Organizational PSM Loss of Containment, fires, explosions, release, severity trend records. South Africa: Organizational Process Safety SHERQ Department.

Restricted. 2012c. Integrated Risk Management Framework-Risk Matrix. South Africa: Process Safety HSE Department.

Restricted. 2011. Organizational Team Leadership Barret Survey Results for Gas-to-Liquid, Effluent and Disposal, Steam Utilities and Ammonia Plants. South Africa: Barret Values Centre.

Restricted. 2010. Process safety second party audit report. South Africa: Process Safety HSE Department.

Restricted 2009a. Organizational Fires, explosions, release severity classification procedure. South Africa: HSE Department.

APPENDIX A PROCESS SAFETY STANDARDS AND INCIDENT ANALYSIS

1. HSE, 2010 and OSHA PSM CFR1910 Process Safety Standard Requirements

Employee Participation

The organization should strive towards "Winning the hearts and minds of employees and sub-contractors" to 'live' process safety. Employees need to participate in the implementation of all the remaining standards and follow safe work procedures. Emphasis is made on teamwork and process safety employee role definition

Process Safety Information (PSI) and Trade Secrets

Describes the requirements of what information is critical and shall be available to all employees to work safely. Employees sign a non-disclosure agreement in the event that sensitive information is distributed. Configuration control procedures and the plants existing quality management system also addresses certain aspects of this standard

Process Hazard Analysis (PHA) and Emergency Response Planning (ERP)

Plant and process risk assessment requirements are described, and employees need to participate or provide technical inputs in all risk assessment exercises. All employees need to be aware of all hazards on the plant as well as the risks created by their neighbors together with the preventative and corrective control measures that shall be taken during emergencies. Emergency action plans and response needs to be compiled and exercised regularly as well as provision of sufficient emergency equipment and maintenance thereof

Standard Operating Procedures (SOP) and Training

This element describes the requirements when compiling operating procedures used in maintenance, plant, and process operations. The training of employees in new or revised procedures are also discusses in the PSM Training standard

Contractors Management

Provides process safety requirements of sub-contractors when they are on site, regarding induction training, entry/exit, understanding unique hazards of each plant, recognizing craft skills and quality requirements for each plant specific operations

Management of Change (MOC) and Pre-Start up Safety Reviews

All plant and organizational changes need to undergo a formal process where the risks arising from the change is fully addressed. When all risks are addressed, a pre-startup safety check is done. MOC is also conducted during organizational restructuring or appointment of key plant managers

Work Permits

A permit is issued whenever work is conducted on plant or machinery. The permit requires a task risk assessment and presence of a safe maker and safety standby during work execution

Mechanical Integrity

Provides maintenance requirements on process safety critical tasks, inspection frequencies identifies selected equipment that is safety critical and provides minimum requirements on risk based maintenance and reliability centered maintenance

Incident Investigation

Provides minimum requirements when conducting an investigation after an incident. The standard makes provision for employee and community incident communication and how incidents are facilitated depending on incident severity

Process Safety Compliance Audits (System and On Site)

Describes how audits are conducted for process safety on three levels Employees within each department conduct the first level, an independent Process Safety Specialist conducts the second and the third is an external audit conducted by an independent and internationally recognized organization

APPENDIX A

Table 1A. Organizational/human factors related to accidents

STORAGE TANK ACCIDENTS	PIPING SYSTEM ACCIDENTS	PROCESS VESSEL ACCIDENTS
1.1 Human and Organizational 36 out of 108 (33%)	1.1 Human and Organizational 41 out of 234 (18%)	Human and Organizational 12 out of 72 (17%)
1.1 Organizational failure, 25 out of 36 (69%)	**1.1 Organizational failure, 26 out of 41 cases (63%)**	**2.1 Organizational failure, (83%)**
1.1.1 Poor planning	1.1.1 Contractor management, 18%	2.1.1 No procedure/system-double/ physical check, 32%
1.1.2 Lack of Analysis	1.1.2 Work permitting, 12%	2.1.2 Lack of analysis, 21%
1.1.3 No procedure/double physical check	1.1.3 Poor management system, 10%	2.1.3 Improper use of equipment, 11%
1.1.4 Improper use of equipment, 10%	1.1.4 No procedure-problem reporting, 8%	2.1.4 Lack of supervision, 11%

1.1.5 Work permitting, 10%	1.1.5 Lack of inspection, 8%	2.1.5 Work permitting, 11%
1.1.6 Lack of Supervision	1.1.6 Poor communication, 8%	2.1.6 Lack of cleaning/maintenance, 5%
1.1.7 Lack of inspection, 6%	1.1.7 Poor planning, 8%	2.1.7 Poor communication, 5%
1.1.8 Lack of maintenance, 6%	1.1.8 Lack of maintenance, 6%	2.1.8 Poor planning, 5%
1.1.9 Contractor management, 4%	1.1.9 Lack of supervision, 6%	**2.2 Human failure, (17%)**
1.1.10 Management of change, 4%	1.1.10 Poor safety culture, 6%	2.2.1 Not follow procedure, 67%
1.1.11 Poor communication, 2%	1.1.11 Improper use of equipment, 4%	2.2.2 Poor training, 33%
1.1.12 Poor safety culture, 2%	1.1.12 Management of change, 4%	
1.2 Human failure, (31%)	1.1.13 Misjudgement, 2%	
1.2.1 Misjudgement, 32%	**1.2 Human failure, (37%)**	
1.2.2 Not follow procedure, 32%	1.2.1 No procedure-double/physical check, 25%	
1.2.3 Knowledge based/ignorance, 21%	1.2.2 Misjudgement, 14%	
1.2.4 Carelessness, 11%	1.2.3 Not follow procedure, 14%	
1.2.5 Poor training, 5%	1.2.4 Poor training, 11%	
	1.2.5 Poor/wrong instruction, 11%	
	1.2.6 Carelessness, 7%	

	1.2.7 Work permitting, 7%	
	1.2.8 Improper use of equipment, 4%	
	1.2.9 Knowledge based/ignorance, 4%	
	1.2.10 Poor management system, 4%	

ORGANIZATIONAL AND HUMAN FACTORS

Source: Data from Foord and Gulland, 2006.

APPENDIX B HUMAN FACTORS SURVEY AND ASSESSMENT

1. Framework for Human Factors Perception Survey

To what extent on a scale of 1 (Least preferred) to 5 (Most preferred) is the following Human Factors Elements adequately implemented and sustained in your work area. Please tick the appropriate box

Table 1B. Human factors perception survey

Human Factors Elements	1	2	3	4	5
Competence and Training					
Human Factors & Risk Assessment					
Incident Reporting and Investigation					
Procedures					
Alarm Handling					
Maintenance					
Behavioral Safety					
Safety Critical Communications					
Control Room Design & Interfaces					
Staffing and Workload					
Change Management					

Process Safety Management					
Supervision					
Leadership					

2. Human Factors Recommendations and Interview Assessment

Table 2B. Human factors summary recommendations and findings template

Human Factor Category	Recommendations			Person Accountable	Due Date	Budget
	Accepted [Yes/No]	Rejected State Why?	Modified			
Maintenance Work						
Emergency Equipment: Maintenance Error						
Field Display, Labeling and Equipment Operation						
Control Room						
Alarm Handling						
Process Control System						
Safety Systems						
Procedures						
Safety Critical Communication						
Remote Operations Communications						
Labeling						

Human Factor Category	Recommendations			Person Account-able	Due Date	Budget
	Accepted [Yes/No]	Rejected State Why?	Modified			
Shift work Is-sues						
Safety Culture						

Table 3B. Human factors template for modern energy industries. Adapted From HSE, 2009.

No	Question	Site Response	Improvements Needed
Maintenance			
1	Is there adequate access for routine operation and mainte-nance of equipment?		
2	Is equipment easily found and recognized?		
3a	Are critical maintenance tasks identified?		
4	Are self-checks/checklists used during maintenance ac-tivities?		
5	Are post-maintenance **checks** performed to detect errors in addition to Pre-Start Up Safety Review?		
6	How is maintenance work pri-oritized? Discuss: Role of sub-contrac-tors, maintenance resources, commitment by sub-contrac-tors regarding priority of maintenance work		
7	Is there a link between prevent-ing loss of containment and		

No	Question	Site Response	Improvements Needed
	general plant/equipment reliability? Discuss: Mean time between failures and management thereof		
8	**Resource allocation:** Is there an adequate system for maintenance resourcing, planning and prioritization?		

No	Question	Site Response	Improvements Needed
Emergency Equipment: Maintenance Error			
1	Is there evidence that maintenance is firmly based on a robust understanding of, and linked to, an analysis of the site's major accident hazards? • Are safety-related & safety-critical maintenance items and activities reliably identified? • Are associated job aids and procedures developed for these priority items? • Is human failure, including violations and error, understood and addressed/managed?		

No	Question	Site Response	Improvements Needed
2	**Formal communication:** Are major accident hazard safety requirements and priorities communicated regularly and reliably to key staff?		
3	**Work design:** Is attention paid to design of maintenance tasks? • How is critical work scheduled (e.g. shouldn't be planned for the end of long shifts/cross-shift)? • Is fatigue managed e.g. is overtime monitored individually; are clear limits set on hours?		
4	Is there **adequate access** to critical valves (incl. elevated valves, control valves, drainage points and drain valves), valve manifolds and field instruments?		
5	Can critical valves be closed off of shut off from a safe location in a timely manner?		
6	Are emergency shutdown switches guarded against inadvertent operation (consider location, switch operation, and guards or covers)		

No	Question	Site Response	Improvements Needed
7	Are field instrument indicators routinely checked for accuracy?		

No	Question	Site Response	Improvements Needed
	Field Display, Labeling and Equipment Operation		
1	Are color codes used consistently?		
2	Are remote startup/shutdown switches clearly labeled and protected from inadvertent operation?		
3	Are remote switches for different systems separated by sufficient distance to prevent operation of the wrong system during stressful situations?		
4	Are operating ranges for process variables specified in the same engineering units as the instrument read-out or indicator (i.e., mental conversion of units is avoided)?. Check calibration of field instrumentation (e.g., avoid using a 0-2500 psig pressure gauge on a 100 psig system)?		

No	Question	Site Response	Improvements Needed
	Control Room		
1	Do screens provide only the information the operator needs (not excessive detail) AND is logically presented?		
2	Do screen schematics correlate with the actual plant configuration?		
3	Is the process appropriately subdivided across screens?		
4	Are board-mounted shutdown switches or buttons sufficiently distinguishable and separated from alarm acknowledgment buttons to minimize inadvertent operation? Are control system display targets (touch screens) spaced adequately to prevent accidental operation?		
5	How many screens do you typically, actively use: When the plant is in steady state? When the plant is starting up? When the plant is in abnormal operation? Would larger display screens help?		

6	Do you find it easy to navigate through the screen hierarchy during normal operations AND in abnormal; situations? Any comments?		
7	How many operations (e.g. mouse clicks) does it typically take for you to get the display format you wish to view and do you get lost in the display format hierarchy?		
8	Are the controls and displays arranged logically to match the expectations of the operators?		
9	Does the layout provide adequate access, egress and freedom of movement		

No	Question	Site Response	Improvements Needed
Alarm Handling			
1	Is the system 'context sensitive'? Does it recognize different operational states and the different operator needs e.g. normal/upset/emergency & what has and hasn't occurred?		
2	Are safety-critical alarms clearly **distinguished and separately** displayed (and hard-wired)? Comment: Discuss grouping of alarms		
3	How are the alarms prioritized? <ul><li>Do operators find the categorization appropriate?</li><li>Targets: high priority 5%, medium 15% and low 80%</li><li>Target alarm occurrence rates: safety-critical -very infrequently; high priority –less than 5 per shift; medium priority –less than 2 per hour; low priority – less than 10 per hour</li></ul>		
4	Can the alarm list be filtered e.g. by priority or plant area?		
5	Resetting of alarms should only be possible if *cleared* (i.e. have returned to normal) *and accepted* by operator		

6	Is there an adequate alarm log/history? • What information is recorded? • How is the information used?		
7	Are emergency arrangements adequate? • Are there enough people available at all times (especially out of hours) to cover for emergencies?		
8	Is the cause of nuisance alarms determined and repaired in a timely manner?		
9	Are return-to-normal indications provided? Discuss (if applicable): Alarm flooding, cascade alarm suppression		

No	Question	Site Response	Improvements Needed
Process Control Systems			
1	Can operators force control inputs into a desired state?		
2	Does the computer check that values entered by operators are within a valid range?		.
3	Are automatic features provided when a process upset/condition may be difficult to diagnose in a timely manner due to complicated processing of information (requiring a knowledge-based decision)?		
4	Can the operator determine the current status of the system versus the desired state?		
5	Is there a dedicated Emergency Shutdown panel and is it located on an egress route?		
Safety Systems			
1	Is there adequate margin between safety system set points and normal conditions?		
2	Is there a documented procedure for bypassing safety systems?		
3	Is a log of bypassed interlocks kept?		
4	Is initiation of safety systems automatic?		

No	Question	Site Response	Improvements Needed
Procedures			
1	For duplicate processes, are the procedures complete and accurate for each process?		
2	Does the written procedure match the way the task is done in practice?		
3	If more than one person is required to perform the procedure, is the person responsible for performing each step identified?		
4	If multiple actions are included in a single step, can the actions actually be performed simultaneously or as a single action?		
5	Does the procedure provide instructions for all reasonable contingencies?		
6	Do procedures that specify alignment such as valve positions, pipe and spool configurations, or hose station hook-ups: specify each item, identify each item with a unique number or designator, specify the position in which the item is to be placed, and indicate where the user records the position, if applicable?		
7	Do maintenance procedures include required follow-up actions		

No	Question	Site Response	Improvements Needed
	or tests and tell the user who must be notified? Are maintenance **checklists** available in each procedure?		
8	Are the types (**checklists,** instructions, flow sheets etc.) of procedure appropriate for: Routine operations? • Safety critical operations? • Emergency and upset conditions?		
9	Is there an ongoing monitoring system to ensure compliance to procedures? (Slips, violations and mistakes) Do the results of this monitoring feed back into the review/revision/validation process?		

No	Question	Site Response	Improvements Needed
Safety Critical Communication			
1	Is there a defined structure for shift handover arrangements?		
2	Do operators know when and how to report safety concerns?		
3	Is there evidence that changes to practices, as a result of an incident, are understood by staff?		

No	Question	Site Response	Improvements Needed
4	Does a process exist to monitor the effectiveness of the communication of major hazard information? (after an incident?)		
Remote Operations Communications			
1	Is there adequate communication and co-ordination with field operators?		
2	Do workers in remote control rooms have adequate displays and see and hear the equipment they control (camera, microphone)?		
3	Do other activities distract workers at remote locations?		
4	Is the travel time from the control room to the unit acceptable?		
5	Do operators in remote locations periodically spend time in the field?		

No	Question	Site Response	Improvements Needed
Labeling			
1	Are all equipment labels (e.g., vessels, piping, valves, instrumentation, etc.) easy to read (clear and in good condition)?		
2	Are all equipment labels correct and unambiguous?		
3	Are all equipment labels located close to the items that they identify?		
4	Do all equipment labels use standard terminology (e.g. acronyms, abbreviations, equipment tags, etc.)?		
5	Are the equipment labels consistent with nomenclature used in procedures?		
6	Are all components that are mentioned in procedures (e.g. valves) labeled or otherwise identified?		
7	Do switch labels identify discrete positions (e.g., ON or OFF, OPEN or CLOSE)?		
8	Are warning signs (e.g. emergency exit, restricted entry, etc.) clearly visible (consider location and condition)?		·
9	Are the signs easy to read (consider letter size and color)?		.

No	Question	Site Response	Improvements Needed
10	Are labels positioned to minimize wear, installed on a flat unobstructed surface?		
11	Are signs that warn workers of hazardous materials, clean-up, maintenance areas adequately visible and clearly understood?		
12	Do the labels correspond to the procedures and drawings?		

No	Question	Site Response	Improvements Needed
Shift work Issues			
1	Have the effects of shift duration and rotation been considered in establishing workloads?		
2	Is the number of hours personnel work during startup, turnarounds, or high production periods limited so that worker safety and performance are not adversely affected? **Are there adequate safety personnel available during turnarounds?**		
3	Are the operators only in the control room or do they do other things? Discuss: Read emails, make phone calls, run errands		
4	Can operators perform all manual adjustments required during normal and emergency operations (not an excessive number of adjustments required)?		
5	Are provisions in place to limit the time a worker spends in harsh environments (i.e., too hot, too cold, confined space, etc.)?		
6	Is there a standard form used for communicating information between shifts or between work groups?		

No	Question	Site Response	Improvements Needed
Safety Culture			
1	Are safety systems maintained regularly?		
2	**Management commitment** • Where is safety perceived to be in management's priorities (Senior/middle/1st line)? • How do they show this? • How often are they seen in the workplace? • Do they talk about safety when in the workplace and is this visible to the workforce? • Do they 'walk the talk'? • Do they deal quickly and effectively with safety issues raised? • What balance do their actions show between safety and production? • Are management trusted over safety?		
3	Is management visibly involved and committed to safety? Discuss: Management walk about inspections. Do they only emphasis occupational safety instead of process safety?		

No	Question	Site Response	Improvements Needed
	Do you feel comfortable/confident when management asks you questions during inspections or during an incident investigation?		
4	Are there frequent communications about safety and free sharing of lessons learned?		
5	Are there adequate health and safety resources?		
6	Is there a high level of trust between management and the front-line workers?		
7	Are workers empowered to stop work if they feel unsafe?		
8	Are operators trained to shut down the process when in doubt about whether it can continue to operate safely/properly?		
9	Are diverse teams involved in evaluating risks? Do you make use of only one risk facilitator?		

APPENDIX C HUMAN FACTORS BOW TIE (RISK MOLECULES)

Table 1C: Maintenance Human Factors Risk Molecule

Risk: Injury to employees, damage to equipment and operation of incorrect equipment due to inadequate: maintenance checklists, maintenance access and no labeling of process safety critical equipment or identification.

Inherent Risk: Impact: 5 Probability: P5 Level: 2

Preventative Control: Employees should compile updated maintenance checklists and comply to existing procedures for lifting of tools and conducting risk assessments for working in areas of minimal access

Corrective Control: Ergonomic Specialists to conduct work-task-environment assessment and Plant Managers to address concerns related to ergonomics and maintenance access

Residual Risk: Impact: 5 Probability: P3 Level:4

Table 2C: Process-Alarms- Safety Systems Human Factors Risk Molecule

Risk: Inconsistent benchmarking on control room screen layout and parameter confirmation, inadequate review of alarm history and unauthorized trip bypasses that may cause operator to lose process.

Inherent Risk: Impact: 6 Probability: P3 Level: 3

Preventative Control: Compile procedures for control room panel layout and process parameter operations as well as reviewing alarm log history. Provide change management on bypass trip procedure

Corrective Control: Review operating procedures to include trouble shooting guide and not heavily rely on operator knowledge-based decision making. Ensure field and operator employees can easily shut down equipment or increase process switching time duration

Residual Risk: Impact: 5 Probability: P2 Level: 4

Table 3C: Equipment Labeling Human Factors Risk Molecule

Risk: Inconsistent coloring of pipes and 'touch and tag' practises throughout multiple sites and inadequate labeling/coloring of equipment, valves and switches and no reference to SOPs, which can easily cause operational mistakes and requires operators to commit to memory the location and equipment operating characteristics

Inherent Risk: Impact: 7 Probability: 4 Level: 1

Preventative Control: Label all equipment and reference in SOP, including switch positions, labeling of critical valves and pipes as well as consistent color coding across sites. Operators should not be confused regarding location of identical equipment. Compile procedure for labeling of equipment and installation practices

Corrective Control: All process safety critical equipment should be color coded and referenced in SOPs. Positive 'touch and tag' should be included in SOPs and at least two operators at any one time should confirm equipment location before operation

Residual Risk: Impact : 7 Probability: 2 Level: 3

Table 4C: Staffing Levels and Shift Work Human Factors Risk Molecule

---

Risk: Inadequate number of people on shift to complete all tasks including issuing of permits in a short space of time, which may cause slips, violations, lapses or mistakes during process upsets or emergency situations.

Inherent Risk: Impact: 4 Probability: 5 Level: 3

Preventative Control: Conduct work time studies and determine adequacy of manpower resource utilization. Increase plant automation to reduce operator workload and fatigue, and do not allow operators to answer non work-related phone calls or emails. Appoint additional staff during peak times

Corrective Control: Revise operating procedures and ensure two or more operators are utilized during peak times or when permits are issued and plant operational requirements need to be met.

Residual Risk: Impact: 4 Probability: 3 Level: 5

Table 5C: Standard Operating Procedures Human Factors Risk Molecule

---

Risk: Standard Operating Procedures (SOPs) not compiled and employees not effectively trained in accordance with OSHA PSM S1.4 and S1.5 standards for operational and maintenance work and excludes trouble shooting guides and safety critical equipment checklists and no task observations

Inherent Risk: Impact: 7 Probability: 5 Level: 1

Preventive Control: Compile SOPs according to S1.4 Standard after conducting critical task observations and develop management system for reviewing SOPs, trouble-shooting guides and critical equipment checklists. Provide employee training in compliance to competency assurance framework and declare them competent before closing off management of change (MOC) projects

Corrective Control: Supervisor to conduct routine inspections on task execution and monitor and correct slips, lapses, violations and mistakes.

Ensure MOC is not closed off before SOPs are updated and employees are trained

Residual Risk: Impact: 7 Probability: 3 Level: 3

Table 6C: Safety Communication Human Factors Risk Molecule

Risk: Ineffective communication devices (e.g. fixed no zoom functionality CCTVs and single channel two-way radios) that may adversely impact response times during emergency situations or provide inaccurate information during normal operations

Inherent Risk: Impact: 5 Probability: 4 Level: 3

Preventative Control: Conduct communication design review and select CCTV based on Major Hazards and high-risk areas including two-way radio with multiple channels. Investigate sufficiency of two-way radios in each department

Corrective Control: Provide adequate manpower resources to monitor on-site activities especially where process safety critical equipment is located

Residual Risk: Impact: 5 Probability: 2 Level: 4

Table 7C: Safety Culture and Incident Reporting Human Factors Risk Molecule

---

Risk: Management by fear, lack of trust and blame/shame can cause repeat incidents due to inadequate incident reporting and ineffective organizational learning

Inherent Risk: Impact: 7 Probability: 4 Level: 1

Preventative Control: Conduct values driven transformational leadership workshops and frequent facilitated discussions with Industrial Psychologists to transform behaviors from fear, blame and shame towards transparency, trust, integrity, accountability and respect

Corrective Control: Increase communication and openness with aim of organizational learning whenever process safety incidents occur on a site wide basis

Residual Risk: Impact: 7 Probability: 2 Level: 3

Table 8C. Organizational Risk Matrix

Name of Business								Financial EBIT	Safety & Health	Community & Customers	Environment	Gov. Relation	Reputation	Legal	HR	Operations	
Level 3	Level 3	Level 3	Level 1	Level 1	Level 1	Level 1	7	>500m	More than 1 fatalities	More than one fatality	Irreversable impact prestine environment	Breakdown in relations with President and/or ministers	Prolonged international & national condemnation	significant business interruption	International strike action	total loss off production	
Level 3	Level 3	Level 3	Level 2	Level 2	Level 2	Level 1	6	100 - 500m	One fatality	One fatality	Serious national environmental impact	Breakdown in relations with Premier and/or provincial MEC's	International & national criticism	Loose plant operating permit	National strike action	future operations untenable	
Level 4	Level 4	Level 4	Level 3	Level 2	Level 2	Level 2	5	10 - 100m	Permanent Disability	Hospital or multiple press articles	Very serious long term environmental impact at regional level	Breakdown in relations with key people in Gov depts (provincial & central)	Serious negative national criticism	May loose plant operating permit	Strikes at several facilities	Future operations at site seriously affected 6 month loss	
Level 5	Level 5	Level 5	Level 3	Level 3	Level 3	Level 3	4	1m - 10m	Hospitalisation	Press article regarding complaints	Serious but reversible short term impact at regional level	Breakdown in relations at local government level	Adverse national media public attention	severe legal fines	Strike at one facility	Major damage to facility, Prod loss < 6 months	
Level 6	Level 6	Level 6	Level 5	Level 5	Level 5	Level 4	3	100 000 - 1m	LVDC	Complaint regarding eg smell	Moderate reversible short term impact at local level	.	Local attention from media / NGO / public	legal fines	Disputes marches organised stay aways	Moderate damage to equipment/ facility. Prod loss < 1 week	
Level 6	Level 6	Level 6	Level 6	Level 6	Level 6	Level 5	2	10 000 - 100 000	Medical treatment / Restriction	None	Minor effects extending beyond boundaries of installation	.	Minor adverse local / public / media attention & complaints	reportable incident	Grievance	Minor/ Superficial damage to equipment. No production loss	
Level 6	Level 6	Level 6	Level 6	Level 6	Level 6	Level 6	1	<10 000	First aid / no injury	None	Limited impact within plant boundaries	.	Public concern restricted to local complaints	none	Complaints dissatisfaction amongst workers	None	
1	2	3	4	5	6	7											
P1 Unforeseen	P2 Highly unlikely	P3 Very Unlikely	P4 Low	P5 Possible	P6 Likely	P7 Almost Certain											
0 - 0.1%	0.1 - 1%	1 - 5%	5 - 15%	15 - 40%	40 - 75%	75 - 100%											
Probability																	

Consequence

LEGEND

Level 1 Risk	Level 2 Risk	Level 3 Risk	Level 4 Risk	Level 5 Risk	Level 6 Risk

Source: Data from Restricted, 2012c.

www.ingramcontent.com/pod-product-compliance
Lightning Source LLC
Chambersburg PA
CBHW040856210326
41597CB00029B/4866